Veterinary Gui… Rearing Dairy H…

A practical and easy to use guide

By Oliver Tilling

About the author

Oliver Tilling is a veterinary surgeon and Director at Shepton Veterinary Group. The practice is based at Shepton Mallet in Somerset and is responsible for 25,000 dairy cows on 120 farms across Somerset, Dorset and Wiltshire.

Oliver has had a particular interest in youngstock work since qualifying in 2006. He works with practice clients in dealing with disease outbreaks, right through to developing programmes that encompass the entire rearing process from birth to calving. He has developed the practice's youngstock service – providing regular monitoring and input into the rearing of dairy heifers.

Oliver has worked with several industry stakeholders, including AHDB Dairy and AHDB Beef and lamb, several pharmaceutical companies and numerous large calf-rearing organisations. He has written extensively for both the veterinary and farming press and travelled to Europe and North America with his work.

Oliver lives in Wells, Somerset with his wife Leanne, son Rupert, two cats Henry and Stanley and his wife's horse Littleman.

Acknowledgements

The author would like to thank the following.

Context Publishing for all their help in publishing this book. All the directors and staff at Shepton Veterinary Group for their continued support and encouragement. The farm clients of Shepton Veterinary Group for their help, patience and fantastic knowledge of rearing dairy heifers. Peter Edmondson for his enthusiasm and encouragement to undertake this project. For the provision of some of the images in this book: Mick Millar, School of Veterinary Science, University of Bristol; Paul Williams of MSD Animal Health; Colin Penny; and VetLearn, publishers of *Diseases of Cattle in Australasia*. My Mum and Dad, who have supported and helped me in everything I do, and my incredible wife Leanne and son Rupert, who had to endure the long process of producing this book but stuck by me throughout.

Thank You.

Bibliographic details

British Library Cataloguing in Publication Data
Veterinary Guide To Rearing Dairy Heifers
A practical and easy to use guide

O. Tilling.
ISBN 9781899043750

For more information on content/references please contact the author.

Context Products Ltd, 53 Mill Street, Packington, Ashby de la Zouch, Leicestershire. LE65 1WN England, Tel: +44 (0) 1530 411337
Email: info@contextbookshop.com
www.ContextBookshop.com

About the Book
Modern dairy farming is an exciting challenge, one that requires dedicated focused people to ensure the health, welfare and productivity of their cows. The modern dairy farmer has to concentrate on today's cows, but also have one eye on the future in the form of replacement heifers. Rearing dairy heifers is an investment in the future but also the second highest cost on a dairy farm. The need to do it well is crucial. This book will guide you through the crucial steps of rearing dairy heifers so as to maximise success and have a productive future.

Veterinary Guide to Rearing Dairy Heifers focuses on 7 key areas that will each give you the essential skills for success:

About Context
Context Publishing is a specialist publisher based in England producing targeted practical technical publications. Our expertise allows us to plan, design and produce publications that are colourful and easy to read yet providing you with the essential information you need quickly. We can also offer versions in different languages if required.

Calling all potential authors!
We are always interested in new books in new areas so please get in touch.

Dr Wesley Ewing

9 781899 043750 >

Sections

1. Calving
- How to be best prepared
- Minimise losses
- What to do in difficult situations and when to call for help
- The first things to take care of in a new-born calf

2. Colostrum
- Understand why this aspect of rearing dairy heifers is so crucial
- Maximise the benefits of good colostrum management and monitor success

3. Feeding from birth to weaning
- Understand the development from milk-fed calf to ruminant and the importance of getting this right
- The options, and which milk and starter feeding programmes to adopt on your dairy
- Achieving a successful weaning transition

4. Feeding from weaning to calving
- How to capitalise on pre-weaning nutrition and achieve growth targets at 6 months, 12 months, service and age at first calving

5. Housing
- Understand the pros and cons of housing calves individually vs groups
- What housing is appropriate for older heifers
- Managing housing systems by focusing on the key areas of moisture, fresh air, air speed, temperature, light, space requirements for feed and water, and cleanliness and hygiene

6. Getting the heifer in calf
- Why to focus on calving heifers at 24 months
- What good reproductive programmes look like and the options for achieving them

7. Dealing with disease
- Diagnosing, treating and preventing common disease problems in dairy heifers from scour and pneumonia through to internal and external parasites

Calving

1

Being born is one of the most perilous moments in any animal's life (Fig. 1). On average 8.1% of calves are dead within 48 hours – a stillbirth. Of this figure 78.6% are born dead – although most are alive when calving starts, and 21.4% are born alive – but die within 48 hours.[1] Keeping these figures as low as possible is a must as birth is likely to represent the period when most calves die on a farm. For calves that survive but have difficulties around calving, there are increased odds for failure of passive transfer (absorption of antibodies during the first 24 hours after birth), and disease and death later in life.

Preparation for calving

Move the cow from the close-up dry cow area into the calving pen as close to calving as possible, but no later than 24 hours before calving. If she spends too long in the calving pen it can get too dirty.[3] The calving area should be clean, dry, with good grip and quiet (Fig. 2). Sand or rubber matting covered in deep straw is ideal. There should be good lighting to facilitate observation, facilities to provide food and water, and the calving pen should be isolated from other areas of the dairy to prevent the cow from becoming distracted or stressed by commotion from other farm tasks.[3] The door needs to be large enough to carry a cow in the bucket of a front-end loader and also to remove the unfortunate fatalities that may occur.[4]

TOP TIPS

Good calving management to improve calf survival and health requires:

- Provision of a suitable calving site
- Adequate but not intrusive calving supervision
- Correct calving techniques
- Judicious utilisation of veterinary assistance[2]

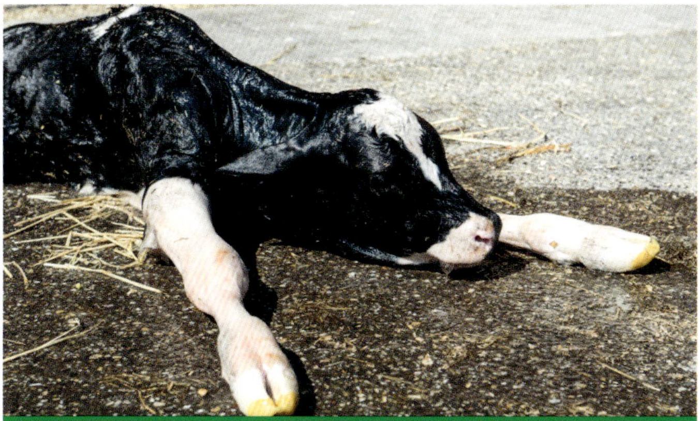

Fig. 1: New Born Calf. The first 48 hours are critical

Fig. 2: Quiet, clean and dry calving area with good grip

Calving

The width of an individual calving pen must be at least 1.5 times the cow's length, and its length should be at least twice the cow's length. A Holstein Friesian cow is 2.50 m–2.75 m from head to tail so the minimum size of calving pen should be 4.2 x 5.5 m (Fig. 3). This should be adequate to allow the cow to lie down comfortably and deliver a calf. If assistance is required, facilities need to allow for at least one cow length behind the cow.[5]

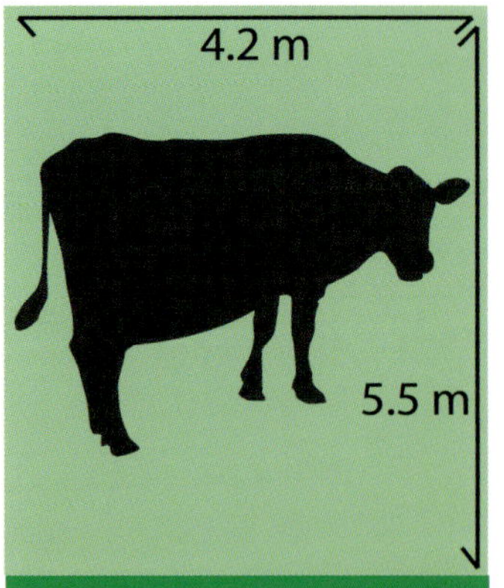

Fig. 3: Diagram of calving pen dimensions

Clean and disinfect calving pens thoroughly between uses to prevent build-up and spread of pathogens. Individual calving pens that can be cleaned between uses are ideal for the prevention of disease. One per 25 cows is required if calving is all year round.[3]

Group calving pens, if managed well, can also be effective (Fig.4). Group pens require fewer workers for monitoring but cows often spend longer in them, leading to a greater build-up of pathogens. Each cow needs 15 m^2 of space and 8-12 kg of straw per day. House less than 10 cows to prevent overcrowding and to prevent lack of feed space, reduced dry matter intake and reduced subsequent reproductive efficiency.[3] Never house sick or lame cows in the same pen as calving cows – they represent a huge risk to the newborn calf.

Fig. 4: Group calving pen

TOP TIPS

Calving pen size

- Width – minimum 1.5 x animal's length
- Length – minimum 2.0 x animal's length

Calving pen numbers

- One per 25 cows

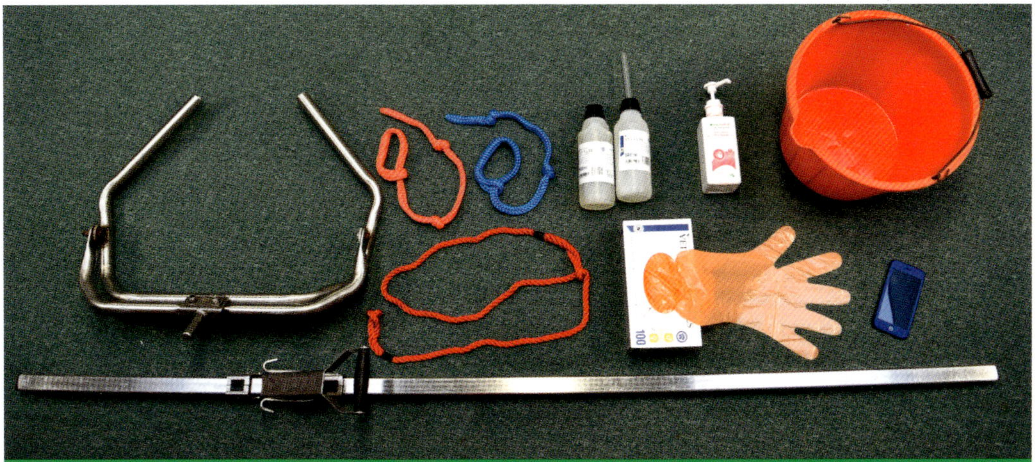

Fig. 5: A complete calving toolkit

The final thing to ensure before a calving is that all necessary equipment is available, clean, sterilised and in good working order. This should include a calving aid, obstetrical chains or nylon obstetrical straps, a head rope, plenty of obstetrical lubrication, long sleeved disposable gloves, 10% iodine and a supply of frozen colostrum if required. There should be clean buckets available, with access to warm water and soap or antiseptic solution. Useful telephone numbers must be easily accessible should further help be needed (Fig. 5).

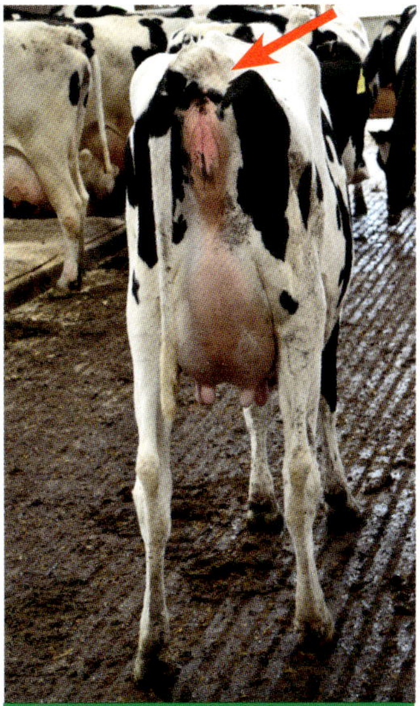

Fig. 6: Udder, tail and vulva at calving. The udder tissue becomes more active at calving, which makes the udder bigger. It becomes full with colostrum and the teats lose their wrinkled appearance. Ligaments around the tail head relax (red arrow). Mucous discharge from the vulva increases and it becomes swollen.

Calving

Practically, time sequences involved in calving are more important than the biological processes. The average gestation of a Holstein Friesian cow is 280 days, but it can vary from 275 to 285 days. Timing of calving is difficult to predict on the basis of clinical signs – but the following should be looked for to give the best indication:

- The first indication is udder enlargement, which usually begins 2–3 weeks before calving in cows and earlier in heifers
- 1–4 days before calving the teats become full, and lose their wrinkled appearance, as the udder fills with colostrum
- At the same time, the ligaments and muscles on both sides of the tail head relax, giving a sunken or hollow appearance to the tail head region
- Mucous discharge from the vulva increases, and it becomes swollen (Fig. 6)[6]
- Given the opportunity the cow will wander off to a secluded area, rumination frequency decreases and she stops eating[4]

Although normal calving is a continuous process, it is often described in three stages. Dystocia (abnormal or difficult calving) occurs when any stage is slow in developing or fails to progress normally.

CALVING STAGE 1

Signs of labour may be absent in cows, but more evident in heifers, where uneasiness and pain with occasional kicking at the belly and flicking the tail may be seen. Restlessness and a tendency to lie down and get up frequently can also be seen. Stage 1 begins with contraction of the longitudinal and circular muscle fibres of the uterus and ends when the cervix is fully dilated and foetal parts enter the birth canal.[7] Uterine contractions first occur about every 15 minutes, but by the end of stage one they occur about every 3 minutes. As the first stage progresses, the contractions become strong enough to cause the cow to arch her back and strain slightly.[7]

Stage 1 lasts 2 to 6 hours, sometimes longer in heifers.

CALVING STAGE 2

The cervix is now fully dilated and the second water sac (amniotic sac), plus foetal parts enter the birth canal further stimulating stronger uterine contractions. The unbroken water sac is often forced through the vulva after the cow has been in labour a short time (Fig. 7).[7] This is probably the most practical indication the animal is in second stage labour. When pressure is applied to the birth canal by the foetal parts, abdominal straining accompanies uterine contractions. The pain of uterine contractions at this point usually force most cows to lie down. Abdominal straining is more frequent as labour progresses until it occurs up to 1–3 times per minute.[7]

The most frequent calf delivery is a frontward presentation, with the calf lying on its front and both front legs and the head extended into the birth canal (Fig. 8). Sometimes, a backward presentation may occur and be deliverable if the calf is the right way up, and both hind limbs are in the birth canal. All other presentations are abnormal.

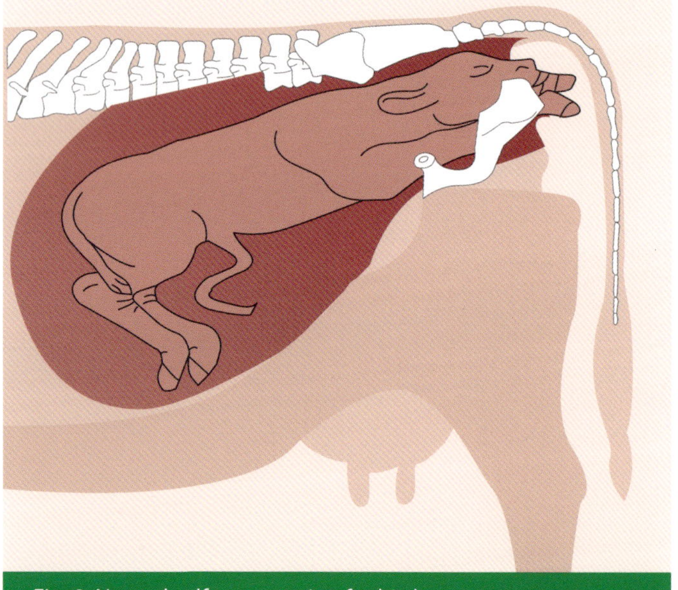

Fig. 8: Normal calf presentation for birth

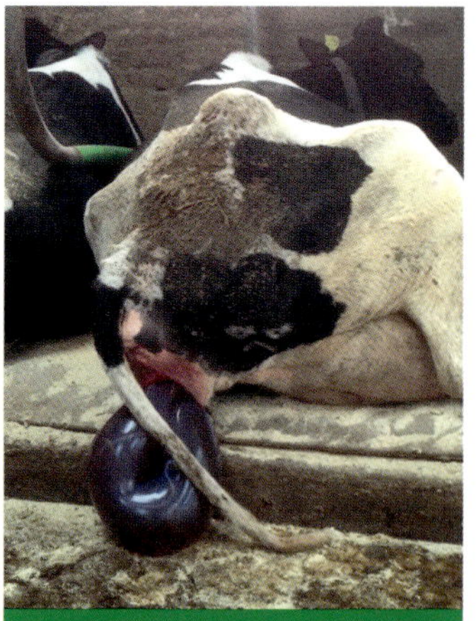

Fig. 7: The unbroken water sac forced through the vulva

During delivery, frequent abdominal straining followed by short periods of rest is normal. The greatest frequency and force occurs when the head is being forced through the birth canal and vulva. Following delivery of the head, the cow may rest for a short time. Strong expulsive efforts are required again to force the shoulders and chest of the calf through the birth canal. The cow may stop straining for a short time following delivery of the chest, allowing the rear legs to rest in the birth canal. At this point, the umbilical cord may be compressed shutting off the oxygen supply to the calf from the cow, so the calf may start to establish its own breathing. Delivery of the hips and legs occurs soon after the chest passes through the vulva and is usually uneventful.[7] Stage 2 lasts 30 minutes to 4 hours.

CALVING STAGE 3

The placenta or foetal membranes are usually expelled within 6 hours after delivery of the calf and the reproductive tract begins to reduce in size.

STAGES OF LABOUR	
1	Coordinated contractions of the uterus
	Positioning of the foetus for birth
	Dilation of the cervix
	Dilation and relaxation of vagina and vulva
	Time: 2–6 hours (up to 8 hours in heifers)
2	Abdominal contractions add to the expulsive forces to deliver the foetus
	Successful delivery of the foetus
	Time: 30minutes–4 hours
3	Release and expulsion of foetal membranes
	Early contraction in size of the reproductive tract
	Time: <6 hours

Table. 1: The stages of labour in the cow

When to assist a calving?

Assistance is bad for the calf as it reduces its survival and increases injuries and infections.

Assistance is bad for the cow as it reduces her production and fertility, while increasing her risk for metabolic problems.

Assistance is bad for people as it can be stressful and reduce morale!

Assistance should therefore only ever occur when it is needed, and never to speed up a delivery. In determining whether to assist, there are several decision-making processes that can have a significant effect on the outcome of a calving.

1. Frequency of observation

Monitoring every 3 hours from the first detection of the onset of stage 1, will allow detection of the onset of stage 2 of normal calving and/or abnormal calving early.

2. When to intervene

Stage 1

If the cow has been in stage one for over 8 hours, intervention is indicated. Abnormal deliveries may not allow the cow to progress, or the cow may be in a state of uterine inertia (an absence of effective contractions) and will not go into stage 2 of labour.

Stage 2

- If the water sac is visible for 2 hours and the cow is not trying
- If the cow has been trying for over 30 minutes and has made no progress
- If the cow has stopped trying for over 15–20 minutes after a period of progress. Breaks normally should not exceed 5–10 minutes unless fatigue or uterine inertia is involved
- If the cow or calf is showing signs of excessive fatigue and stress, such as swollen tongue of the calf or severe bleeding from the rectum of the cow
- If an abnormal delivery is determined[7]

Stage 3

If the cow has not passed foetal membranes within 12 hours of calving, intervention may be necessary. If they are retained, treatment may be indicated. However, do not manually remove foetal membranes as this is detrimental to subsequent reproductive performance[7].

3. Can the calf be delivered by forced extraction?

If the decision to intervene is taken, it is important to act in a prompt and patient manner and basic guidelines should be followed. Facilities for assisted delivery should allow easy entry of the cow and minimise the stress of handling and restraint during assistance (Fig. 9). Alternatively, the use of a halter and gates may be necessary, but carries increased risk to those involved. Crushes should be avoided – if the cow goes down during the delivery in a crush it can be a disaster!

Fig. 9: Excellent facilities for calving – a locking yolk with a gate that allows the cow to be guided in and restrained, but can open up if the cow goes down, or access is required for a caesarean

Fig. 10: Nylon straps

A cement floor is recommended for cleaning purposes with access to both hot and cold water. Obstetrical chains are preferred to rope when traction is required as they are more easily cleaned. Also, handles are available that attach anywhere along the chains and make traction easier to apply. Nylon web obstetrical straps are available. These may be less traumatic to the foetus, but must be cleaned very carefully between deliveries (Fig. 10).[7]

Clean hands and arms thoroughly with soap or antiseptic and water. Ask an assistant to hold the cow's tail to one side. Clean the cow's perineal area with soap or antiseptic and water. Insert the hand gently each time upon entering the birth canal, wear long-sleeved disposable gloves and always use plenty of lubrication. At the time of insertion fingers and thumb should be together and pointing forwards, with the thumb uppermost. This will cause the least discomfort to the cow.[4]

- *Once in the vagina push the hand forwards slowly towards the cervix. If the vagina ends in a hard protruding button and in the centre of that button there is a hole into which only one finger can be inserted then the cervix is fully closed and the cow should be left*[4]
- *If the hand can be passed through the cervix, but its rim feels like a ring or a thick fibrous band running around the inside of the vagina, then the cervix is not fully dilated and the cow should be left for a little longer*[4]
- *If the cervix is fully dilated, first determine if the calf is alive. This is usually possible by placing fingers in the mouth to stimulate a suck reflex, or pinching in between the toes of a leg to stimulate a withdrawal reflex*
- *Determine the position and presentation of the calf, whether it is normal or abnormal*
- *If abnormal, determine if correction is possible or further help is needed. Do not pull a calf that is in any position other than normal without first correcting the abnormal presentation*

- *Leg ropes or chains should be tied above the fetlocks of the calf. If a head rope is to be used, it should be placed behind the ears and in the mouth of the calf above the tongue* (Fig. 11)

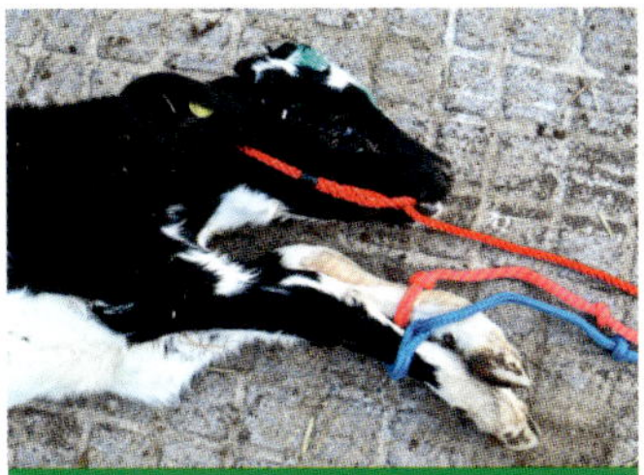

Fig. 11: Correct positioning of leg ropes tied above fetlocks, and head rope placed behind the ears and in the mouth of the calf above the tongue

The lowest numbers of deaths around calving are on farms where access to mechanical assistance is restricted. Mechanical calving aids are useful tools, but if used incorrectly are very dangerous and can cause a great deal of suffering and harm. The main danger is if the calving aid is used in the wrong circumstances, for example, before the cervix or vagina has fully opened, before the calf has been correctly positioned for delivery, or simply when the calf is too large and veterinary assistance is required. Another misuse is that the calf is drawn too quickly. If the tissues of the vulva and vagina are not allowed to dilate naturally as the calf's head is being delivered, tearing of the vagina may occur. This can lead to severe infections or even fatal blood loss.[4]

Anyone who uses a calving aid should have some basic training in how to use one safely.[8] Use may be initiated with the cow standing but there must be room for her to lie down should she decide to, without risk of injury to assistants by the cow herself or equipment being used. Proper use of a calving aid should cause some progress within 1–2 minutes of traction and, if not, then a caesarean may be indicated.

As a general rule the handle on the calving aid should be used to take up the slack in the ropes while traction is applied by gently pulling the shaft down in an arc behind the cow.[8] This pressure should only be applied when the cow is straining as her abdominal muscles contract.[7] This will provide a much more gentle traction than delivering the calf using the handle.[8] Also the calf can only be supplied with oxygen-rich blood between contractions. This blood supply to the calf is at risk if uninterrupted traction is applied.

- Whether using a calving aid or bars placed through the leg ropes, it is important to apply a steady but continual pressure
- Check regularly the head is coming with the feet – if not use a head rope (Fig. 11)
- Apply plenty of lubrication – either a proprietary lubricant or handfuls of dry soap flakes that can be submerged in water and then lathered where most lubrication is needed – the head and shoulders. Failure to provide enough lubrication is a common mistake and can lead to tearing of the vagina
- When the head is passing through the vulva, the calf's ribs will be passing through its mother's pelvis and if the birth is tight the umbilical cord may be constricted. Time is now more important[4]
- Once the calf's head and torso are out, continue to pull, coordinated with the cow's straining, ensuring that the calf's legs are pulled at an angle down towards the cow's feet, rather than straight backwards. This enables the calf to pass in an arc through the cow's pelvis and facilitates passage of the calf's hips[4]
- Additional pressure may be required to get the calf's hips through, and a slight rotation of the calf may also help, so that its hips pass obliquely through the cow's pelvis[4]

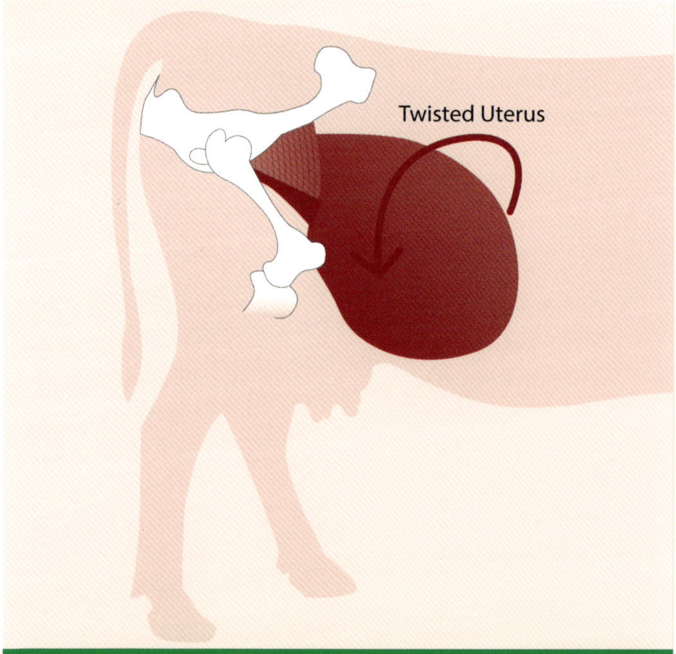

Fig. 12: Diagram representing the position of a uterine torsion

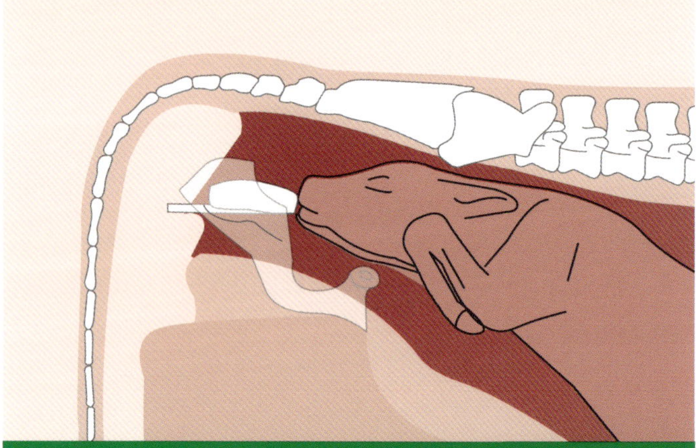

Fig. 13: Leg back flexed at the knee

4. Correcting an abnormal presentation

Experience and confidence are required when correcting abnormalities, as well as being aware of limitations and when to seek further assistance. Ensuring all equipment is available and in good order is essential – especially the provision of copious amounts of lubricant.

Uterine torsion (Fig. 12)
This condition is thought to be caused by an overactive calf at the start of calving that leads to the uterus rotating around its axis. The point at which rotation takes place can be in front of or behind the cervix. Rotations vary between 90º and 270º. Therefore upon examining the cow, the presentation may be one of an abnormally positioned calf that lies at an angle, right through to an abnormal feeling cervix/ inability to feel a cervix. If the hand can be passed through the cervix, the feeling is like that of the rifling inside the barrel of a gun.

It is possible to correct a uterine torsion, but always seek veterinary attention. It is not always possible to correct the torsion and a caesarean section may be required. Sometimes following correction of the torsion the cervix fails to dilate and a caesarean section is still required.

Uterine inertia
If the cervix and vagina are fully dilated, the calf lies in the correct position, but the cow has stopped pushing, it is likely that she has uterine inertia. Essentially, she's run out of energy or more accurately she is low in calcium – a "milk fever". Traction will be required to help the cow deliver the calf. It is also worth administering some calcium to her – preferably intravenously.

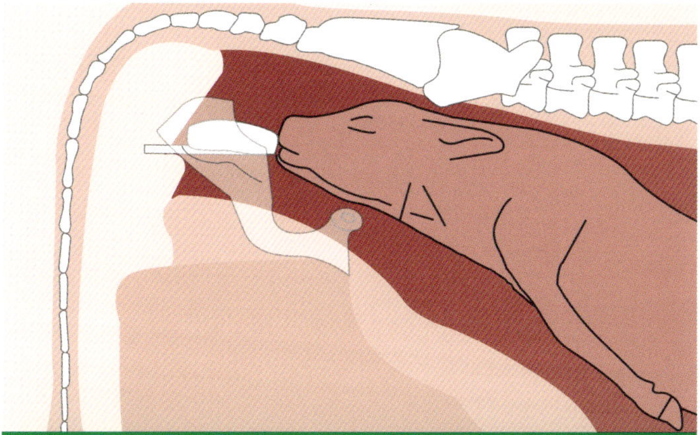

Fig. 14: Leg back flexed at the shoulder

Leg back (Figs. 13–15)
If presented with a head and one leg, or even just a head – then the leg is a back one. The leg can be flexed at the knee or the shoulder. In either situation use the flat of the hand to push the calf's head as far back into the uterus to make space. If the leg is flexed at the knee then using plenty of lubrication the hoof can be cupped in the hand and drawn gently forward into a normal position.

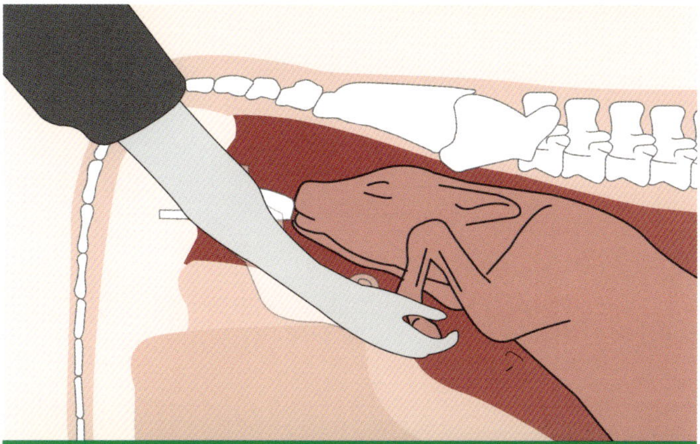

Fig. 15: Correcting a leg back presentation

If the whole leg is turned back, i.e. flexed at the shoulder it is more difficult to correct. With either a long reach or by placing a rope around the back of the knee, apply plenty of lubrication and the leg can be pulled forwards into the "knee-flexed" position and then corrected as described above.

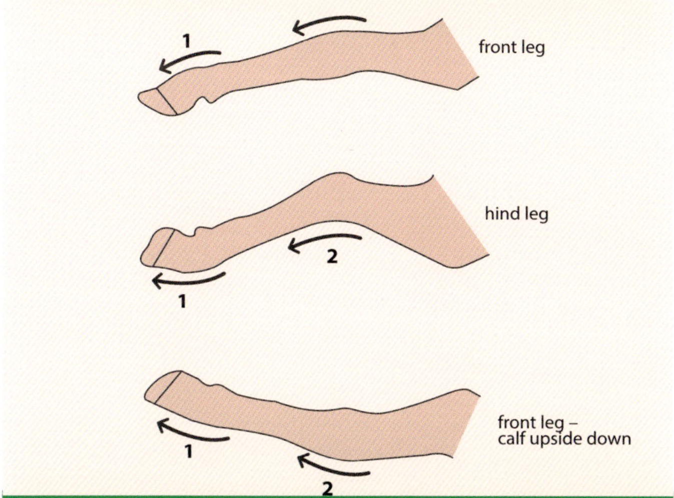

Fig. 16: Determining the difference between front and back legs during calving (From Blowey (1999), A Veterinary Book For Dairy Farmers)[4]

Head back (Fig. 17)
In this case only two legs will be presented at the vagina or outside the cow, but it is essential to determine if they are front or back legs. If the first two joints bend the same way it is a front leg. If the first joint moves the foot up and the second moves the leg down then it is a hind leg (Fig. 16). Be sure they are front legs and then check again!

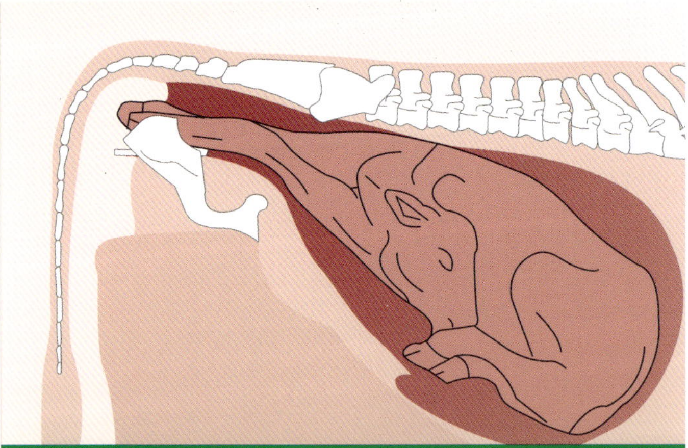

Fig. 17: Head back presentation

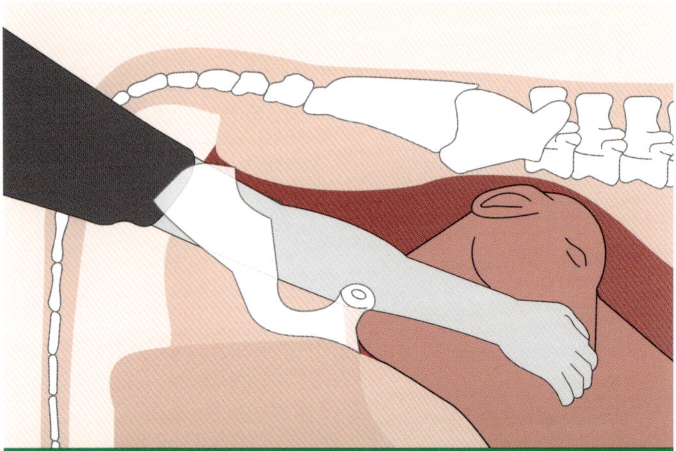

Fig. 18: Correcting a headback presentation

Fig. 19: Lateral pressure to help with a calf stuck at the hips

Place leg ropes on both legs. Follow the legs down, so as to locate the base of the neck and determine the head position. The aim is to pull the head back round by grasping the jaw or pulling on the nose (Fig.18). This may take several attempts, and initially pulling on the neck itself may be required. If excessive force is used then there is the danger of rupturing the uterus so always use plenty of lubrication and go slowly. Once the head is corrected, place a head rope on it to prevent it slipping back when traction is applied.

Hip lock

This occurs during a normal presentation, but the hips of the calf get stuck in the pelvis of the cow – it is very frustrating. It is potentially life threatening to the calf and very damaging to the cow as it can cause severe nerve damage – so move quickly. Options include:

1. Stop, apply lots of lubrication into the cow and try traction again

2. Try to twist the calf at the same time as applying traction – to rotate the hips of the calf to an angle where they can pass through the widest part of the pelvis. This can be done by a second assistant standing to the side of the calf pulling on its chest and pushing on its back (Fig. 19). A rocking motion will often allow the calf to "pop" through. A rope can also be applied laterally to the chest and pulled on repeatedly to the same desired result

3. Lay the cow down – this can often make the difference (Fig. 20)

4. If the calf won't come through and usually by now is dead; call for veterinary assistance – which usually involves an embryotomy (cutting the calf into pieces to remove it)

Fig. 20: Reuff's method for casting a cow, which when laid down will often help with a hip lock

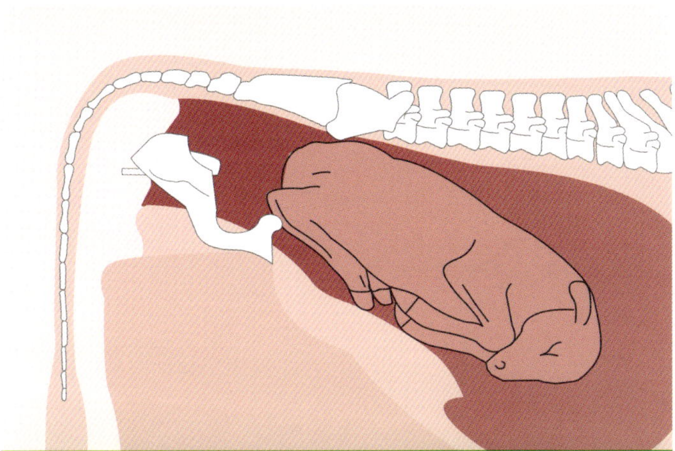

Fig. 21: A calf in a breech presentation

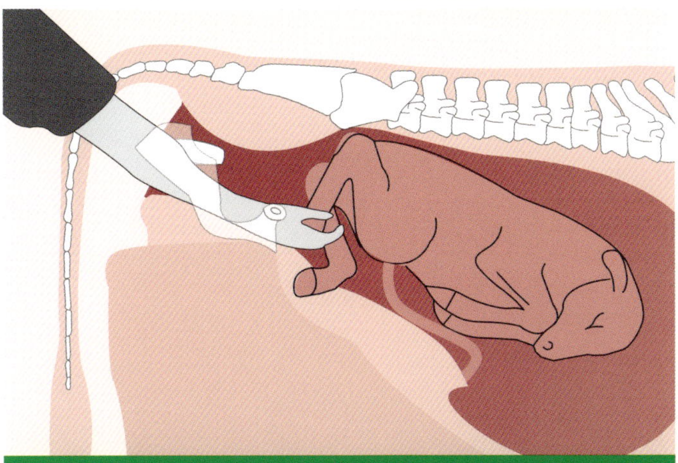

Figs. 22: Correcting a breech presentation

Backwards delivery

Although this is a fairly common delivery it carries a three times greater risk of a stillbirth. This is because the umbilical cord can become pinched early on during calving or tears off and there is a greater risk of amniotic fluid aspiration.

Be certain the legs are back legs (Fig. 16) and then apply ropes. Backwards presentations do not dilate the vagina and cervix as well as a forwards presentation. Ferguson's reflex is the action of the calf's head rubbing against the vagina and cervix of the cow stimulating them to dilate. Without this in a backwards presentation, plenty of lubrication and patience are required to avoid tearing the vagina as the calf's hips pass through. It is also essential that the tail is in its normal position between the calf's legs. If it is pushed up over the calf's back it could cause severe damage to the cow's vagina. Traction needs to be parallel to the cow's spine without pulling down on the calving aid.

Breech presentation

(Figs 21 & 22)

These presentations often require veterinary assistance. Often, all that is presented at the vulva is the tail of the calf. The flat of the hand is used to push the backside of the calf as far as possible back into the uterus, with the aim then to extend the limbs backwards. Choose whichever leg you can get hold of first. Pull the hock into the pelvis (a rope around it may be required). Flex the fetlock by cupping the foot in one hand and then push the hock away with the other hand while, at the same time, pulling the flexed foot towards you. Cupping the foot is very important otherwise it will rupture the uterus. It is made easier by angling the leg inwards slightly, underneath the body of the calf, as you pull it backwards.

Once one leg is done, repeat for the other. Much the same as a backwards presentation, the cervix and vagina are often not fully dilated, so plenty of lubrication and patience are required. A breech is also often a twin – the cause is often thought to be a lack of room preventing each calf assuming a normal position. Always check for a second, which also often presents as a breech!

Transverse presentation (Fig. 23) Call veterinary assistance – a caesarean is usually required.

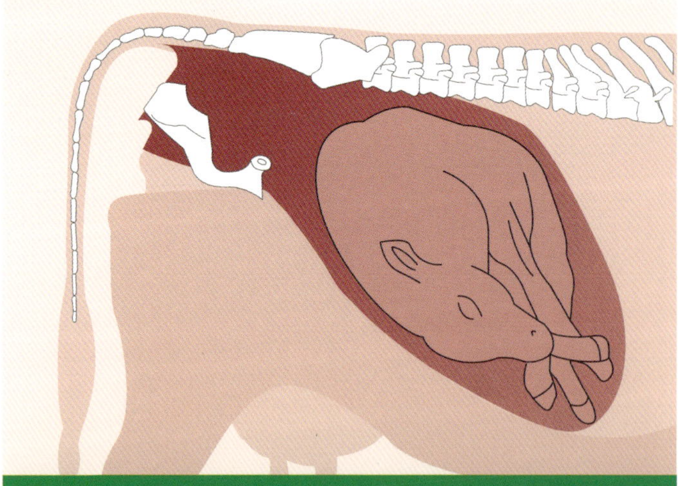

Fig. 23: Transverse position

Monster calves (Figs 24–27) Many times when calving a cow it is possible to be convinced the calf is deformed. Although rare, there are many different abnormalities or monsters. Examples are given in Figs 24–27, but many others exist. Seek veterinary assistance.

Fig. 24: Schistosomus reflexus. A defect in the lower abdomen allows the contents to float free in the uterus

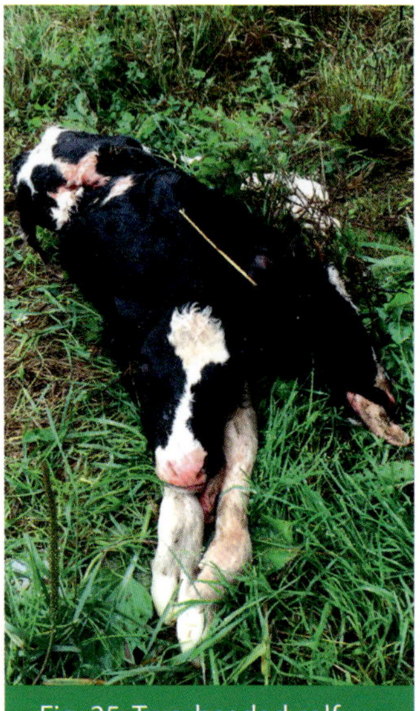

Fig. 25: Two-headed calf

Fig. 26: Amorphous globosus. A hairy ball of tissue with its own organs inside. Most are born twin to a normal calf

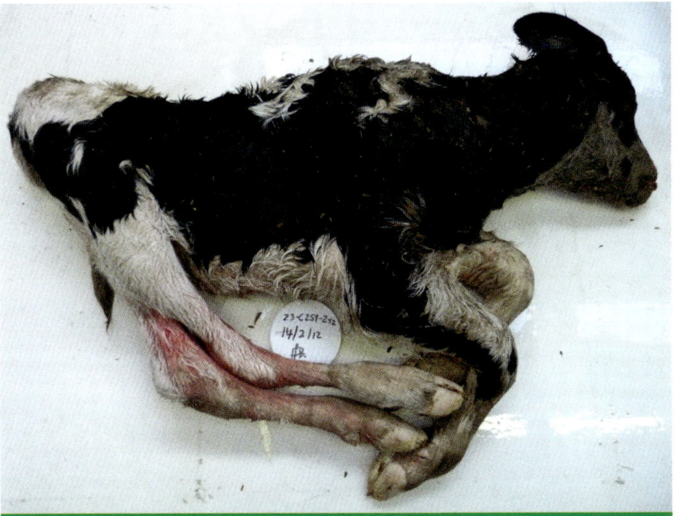

Fig. 27: Arthrogryposis. Joints become fused and difficult to manipulate – a common finding with Schmallenberg virus infection.
(Photo courtesy of M. Millar, School of Veterinary Science, University of Bristol)

5. When to call for professional assistance

- Don't know what problem is
- Know the problem and solution, but is beyond capabilities
- Know the problem and the solution but have tried and simply made no progress in a 30-minute period! Further delays will simply put the calf in jeopardy[7]

Post-calving

After the cow or heifer has calved, the temptation is often to walk away. Table 2 shows normal calf behaviour following birth. If normal behaviour is not observed then resuscitation steps are needed. Of course, if observation is not undertaken then nothing can be done!

A clean, dry towel rubbing along the top line from tail head to head stimulates breathing, then focusing on ears, around the eyes and nose – thus mimicking the actions of the mother. To remove any excess fluid from the lungs position the calf on its chest and head below its tail, i.e. facing down hill. Do not hang a calf over a gate, this is detrimental to it – most of the fluid comes from the abomasum, plus the lungs can be compressed during this action making it harder for the calf to breathe. Pinpoint pressure across the nasal septum is an alternative. However, the best technique is a bucket of ice water on the head of the calf or in its ear – very stimulating! (Fig. 28). Once a patent airway has been established the calf should be placed upright on its chest – allowing both lungs to fully expand. The upper part of the lungs – adjacent to the spine, are larger and so therefore more important.

1. Head righting in minutes
2. Sitting in 5 minutes
3. Attempts to stand within 15 minutes
4. Standing within 1 hour
5. Suckling within 2 hours
6. Heart rate: 100–150 beats/minute
7. Respiratory rate 50–75 breaths/minute

Table 2: Normal Calf Behaviour Following Birth

Calves that experience dystocia but are alive at birth should be given special attention to improve their odds of survival. The calf should be examined carefully for defects or for damage during the birth process. In particular, if the calf was pulled, the legs should be examined for evidence of damage.[6] Calves experiencing a prolonged dystocia are likely to have low levels of oxygen in their blood (hypoxia) and their blood pH is frequently acidic (acidosis) instead of neutral. These impairments lead to a cascade of events, such as decreased ability to nurse, decreased absorption of antibodies and poor temperature regulation. The administration of oxygen to calves after dystocia may improve survival. In addition, careful attention to adequate colostrum intake and maintenance of body temperature are critical.[1]

Fig. 28: A bucket of ice water on the head – very stimulating!

Clearly identify calves that have required extra care following calving by marking them, their pen/hutch and recording their identification clearly in the calf house. This way they can be monitored more closely during the first few weeks of life.

Dip all calf navels at birth with 10% iodine or chlorhexidine. Make sure the navel is fully submerged or sprayed right up to the skin (Fig. 29). Don't use iodine teat dip – it is not the same thing.

Bacterial infection (often *E. coli*) of the navel called "navel ill" or migration of infection to the joints called "joint ill" are associated with poor navel dipping. Navel-dipping calves at birth provides many benefits. Without disinfection, the same lifeline that delivers nutrients and removes wastes while the calf is developing in the uterus becomes a dangerous pathway into the calf's body until complete closure takes place throughout the first 2 months of life.

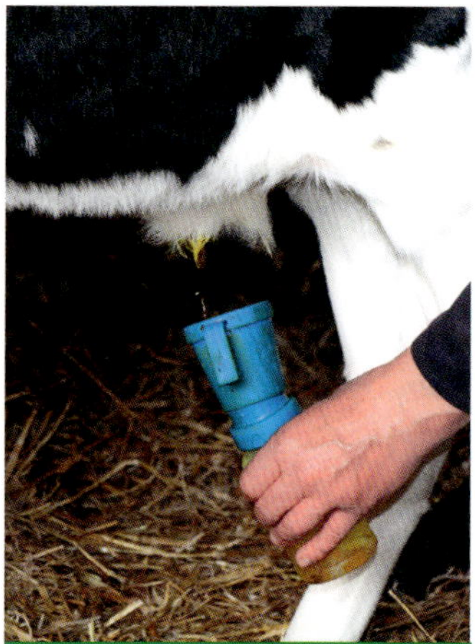

Fig. 29: Dipping a calf's navel with 10% iodine

Calves whose navels are not dipped are twice as likely to die as those whose navels are dipped. Those that survive the easily preventable infection are 2.5 kgs lighter at 3 months of age than their herd mates whose navels were dipped at birth.[9]

Even those farms with well-managed calving areas may have trouble keeping them perfectly clean and dry at all times. Navel-dipping provides calves with a defence against bacteria in the environment that cannot be revisited once skipped at birth. If "navel ill" is a persistent problem, repeat dipping at 24 hours of age, but more importantly, check the cleanliness of the calving areas.

The cow and calving environment are the main source of infectious agents to the calf – these include pathogens that cause scour, pneumonia, "navel ill" and Johne's disease. Remove the calf from the cow immediately after calving. This minimises the risks of infection, and reduces the stress to both the cow and calf.[10] Transport the calf in a clean wheelbarrow or dedicated newborn calf transport vehicle to a clean hutch or pen before administering colostrum.

Don't forget to pay attention to the cow. First, check for a second calf and if you remove a second check for a third. Check for vaginal tears and excessive bleeding and check all four quarters of the udder for mastitis.[4] If the reproductive tract has been torn or damaged – administer broad-spectrum antibiotics. If there is excessive bleeding this needs to be stopped – contact the vet immediately. Any assisted calving must receive pain relief.

Cows and heifers may not eat or drink for up to 24 hours around calving so provide them with a minimum of 40 litres of tepid water and palatable feed. One option is to place the calf on some palatable feed outside the calving area, but where the cow has access to the calf. The cow can then lick the calf clean and stimulate it and when the calf is removed she will readily eat the feed it has been lying on. The freshly calved cow then needs to continue to be monitored. There should be repeated observation during the day for the first 5–7 days after calving monitoring feed intake and behaviour.[8]

It is important to benchmark and monitor calving activity using a simple recording scheme. This can be a sensitive indicator of welfare around calving time. Calving difficulty can be recorded as in Table 3.

A target of less than 10% of cows, and less than 15% of heifers achieving score 3 and 4 should be achievable.[8] Live birth rate should also be monitored and can be defined as the percentage of calves born that are alive and survive for 24 hours. The target would be 97% of cows and 92% of heifers.[8]

Have a standard operating procedure for calving cows, and in this way, everyone who could potentially be involved knows what to do. This must include accurately recording cow and calf numbers each time. A calving is only finished when everything is ready to go for the next one!

CALVING DIFFICULTY	
0	No hand touches calf
1	Hand touches calf but no rope used
2	Rope is used – gentle pull
3	Rope is used – hard pull
4	Caesarean section

Table. 3: Calving score

References

1. NAHMS (2007) *Dairy 2007: Heifer Calf Health and Management Practices on U.S. Dairy Operations*, 2007.
2. Lorenz. I, Mee, J. F., Earley, B. and More, S. J. (2011) Calf health from birth to weaning. 1. General aspects of disease prevention. *Irish Veterinary Journal* 64:10.
3. Mee, J. F. (2008) Newborn dairy calf management. Veterinary clinics of North America, *Food Animal Practice* Vol. 24 No. 1.
4. Blowey, R. W. (1999) *A Veterinary Book for Dairy Farmers* (3rd Edition). Farming Press Figure 5.8, Page 124.
5. Hulsen, J. and Swormink, B. K. (2006) *From calf to heifer. A practical guide for rearing young stock.* Roodbont publishers.
6. Davis, C. L. and Drackley, J. K. (1998) *The Development, Nutrition and Management of the Young Calf.* Ames (IA) Iowa State University Press.
7. Mortimer, R. G. (2009) *Calving Management Manual*, Chapter 6: Calving and handling Calving difficulties. Available at: www.cvmbs.colostate.edu/ilm/projects/neonatal/Calving%20and%20Handling%20Calving%20Difficulties.pdf Accessed January 2015.
8. Fishwick, J. (2011) Welfare issues associated with calving. *Cattle Practice* 19.
9. Virtala, A. M., Mechor, G. D., Grohn Y. T. and Erb, H. N. (1996) The effect of calfhood diseases on growth of female dairy calves during the first 3 months of life in New York State. *Journal of Dairy Science* 79.
10. Albright, J. L. (1987) Dairy animal welfare: current and needed research. *Journal of Dairy Science* 70.

Colostrum

Veterinary Guide To Rearing Dairy Heifers

2

Colostrum

The most important meal in a calf's entire life is its first meal – colostrum. Colostrum is the first thick, yellow milk a cow or heifer produces after calving (Fig. 1). The importance of ensuring the calf receives adequate colostrum cannot be overemphasised. Calves that receive adequate colostrum have a reduced risk for pre-weaning disease and death, with additional long-term benefits including:

- ***Reduced risk of death in the post-weaning period***
- ***Improved rate of gain and feed efficiency***
- ***Reduced age at first calving***
- ***Improved first and second lactation milk production***
- ***Reduced tendency of culling in the first lactation***[1]

Fig. 1: Buckets of freshly harvested colostrum

The main characteristics of colostrum are:

- *It contains antibodies that protect the calf from the wide range of diseases that its mother has been exposed to during her recent life*
- *It is highly nutritious, which is important in getting the calf warmed up and moving around soon after birth and for maintenance and growth*
- *It contains the first water a calf receives*
- *Its increased fat content – as well as an energy source – acts as a laxative and assists in the passage of meconium, the foetal dung*
- *It has very high levels of white blood cells for defence against infections, plus several other factors that stimulate the newborn's immune system (growth factors, hormones, vitamins and minerals)*
- *It stimulates the production of acid and digestive enzymes when it enters the abomasum. At birth the abomasal pH is quite high falling to pH 3–4 at 2–3 days old, when acid is produced. This acid kills many ingested bacteria and is therefore a very important defence mechanism*[2]
- *Hormones and growth factors in colostrum are responsible for determining the expression of certain genes involved with weight gain, mammary development, development of the digestive tract and development of the reproductive tract – this is called epigenetic programming*

Colostrum is very different from milk (Table. 1), being nearly twice as concentrated (23.9% vs 12.5%). There are much higher levels of fat in colostrum compared to milk and much higher levels of protein, especially antibodies (67 times greater in colostrum vs milk).

COLOSTRUM (MILKING POST-CALVING)				
Factor	**1**	**2**	**3**	**Milk**
Total solids (%)	23.9	17.9	14.1	12.5
Fat (%)	6.7	5.4	3.9	3.6
Solids-not-fat (%)	16.7	12.2	9.8	8.6
Total protein (%)	14.0	8.4	5.1	3.2
Casein (%)	4.8	4.3	3.8	2.5
Albumin (%)	0.9	1.1	0.9	0.5
Immunoglobulins (Antibodies) (%)	6.0	4.2	2.4	0.09
Lactose (%)	2.7	3.9	4.4	4.9
Calcium (%)	0.26	0.15	0.15	0.13
Vitamin A (μg/100 ml)	295.0	190.0	113.0	34.0
Vitamin E (μg/g fat)	84.0	76.0	56.0	15.0
Vitamin B12 (μg/100 ml)	4.9	–	2.5	0.6

Table. 1: Characteristics and composition of Holstein colostrum and milk[3]

In late pregnancy the cow concentrates antibodies in her colostrum, so that the calf receives immediate preformed immunity to many of the diseases to which it will be exposed. The final concentration of antibodies in colostrum is much higher than that originally present in the blood and is the reason why the protein content of colostrum is so high.

The cow and calf blood supplies are separated by the cow's placenta within the uterus. This prevents transmission in the uterus of antibodies. The calf therefore depends almost entirely on absorption of antibodies from colostrum after birth[1] (Fig. 2). Absorption of these antibodies across the calf's small intestine, during the first 24 hours after birth, is called passive transfer and helps protect the calf against disease until its own immature immune system becomes functional.

Fig. 2: The newborn calf

The immunity given to the calf is only related to the infections the cow herself has contacted. If a cow is purchased and moved into a new herd only a few days before calving, then clearly there is a risk that the calf will be challenged by infections for which it has no colostral protection.[2] So, ensure all cows and heifers are on a farm at least 3 weeks prior to calving.

Failure of passive transfer of immunity (FPT) occurs when a calf fails to absorb an adequate quantity of antibodies.[4] FPT is not a disease, but a condition that predisposes the newborn calf to the development of disease.[5]

The estimated prevalence of FPT in US dairy heifers is 19.2%,[4] with 31% of deaths in the first 3 weeks due to FPT.[6] A recent study in the South-West of England found 34% of dairy calves had FPT, in other words, 1 in 3 calves had not received adequate colostrum.

THE 5 KEY AREAS TO EFFECTIVE COLOSTRUM MANAGEMENT: THE CORNER STONE OF GOOD HEIFER REARING

Quality
Quantity
Quickly
sQueaky clean
Quantify

Quality

Feed only high-quality colostrum that has been measured using a colostrometer or Brix refractometer.

Good-quality colostrum contains high levels of antibodies. Antibodies are large Y-shaped protein molecules that help the calf fight infection. The concentration of these antibodies in colostrum can vary considerably between individual cows and is highest immediately after calving. Concentration then decreases by 3.7% during each subsequent hour post calving (Fig. 3). Time of first milking is therefore the most crucial factor regarding colostrum quality that farmers can influence.[7]

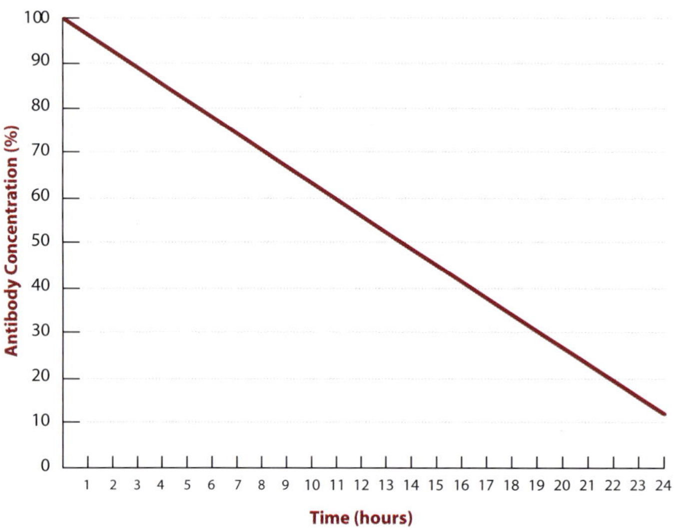

Fig. 3 Concentation of antibodies in colostrum in the first 24 hours after calving

Milk the cow as soon as she calves – and milk her out completely. This first colostrum is the best – **First is Gold.** A cow should produce up to 8 litres of colostrum (a heifer up to 6 litres) and all of that first colostrum should end up in the calf. Colostrum produced at 12 hours post-calving is of significantly poorer quality so the focus must be on the first and best quality. Waiting until the next milking to harvest colostrum is unacceptable.

Other factors under the farmer's control that affect colostrum quality include:

- **Dry cow nutrition** – feeding dry cows and heifers balanced non-lactating rations, with a suitable transition period
- **Vaccination during dry period** – to increase colostral antibodies vs common infections the calf may encounter, e.g. Rotavirus, Coronavirus, *E. coli, Salmonella dublin* and *Salmonella typhimurium*
- **Stressors** – reducing stress the cow is under, for example, heat stress. Reducing stress in the heifer, for example, making sure she is on the farm in a relaxed setting at least 2 months before she is due to calve – with the added benefit she has time to be exposed to pathogens and produce appropriate antibodies for her calf
- **Length of dry period** – cows with excessively long dry periods (>90 days), short dry periods (<21 days) or no dry period, produce colostrum with significantly lower antibody concentrations
- **Do not pool colostrum** – low antibody, high-volume colostrum will be over-represented in any pooled colostrum[5]. The use of pooled colostrum increases the risk of FPT by 2.2 fold[4]

Several other factors will have a bearing on colostrum quality, which are beyond the control of the farmer:

- *Beef breeds produce better quality colostrum than dairy*
- *Older cows having greater exposure to on farm pathogens produce higher quality than young cows*
- *Dilution effects of high yielding cows can lead to poorer quality colostrum – if more than 8.5 litres is obtained the quality is likely to be poor*[5]
- *Sick cows suffering mastitis or milk fever inevitably produce poorer quality colostrum.*

Telling the difference between high - and low - quality colostrum is not easy. Certainly, assessing by eye has caught many a farmer out! The most frequently used instrument is the colostrometer, a hydrometer instrument that measures antibody concentration by measuring colostrum specific gravity. It is colour coded to provide guidance in colostrum selection (Figs. 4–7). Assessment of colostrum quality with the colostrometer can be used as the basis of identifying situations where colostrum supplementation or replacement is required, as well as identifying good-quality colostrum suitable for storage (Fig. 8).[8]

Fig. 4: Colostrometer showing good-quality colostrum

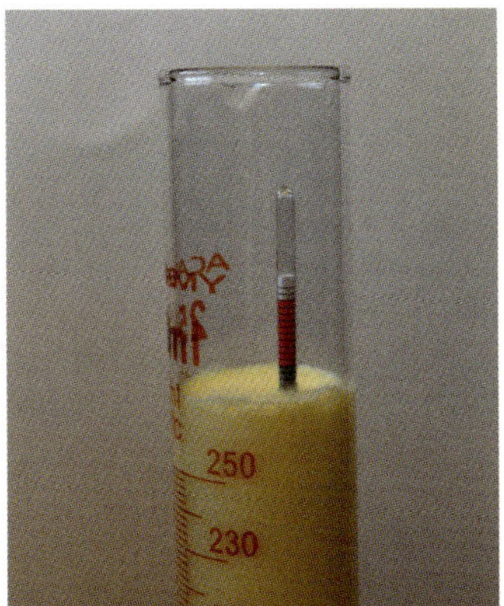

Fig. 5: Colostrometer showing average-quality colostrum

Fig. 6: Colostrometer showing poor-quality colostrum

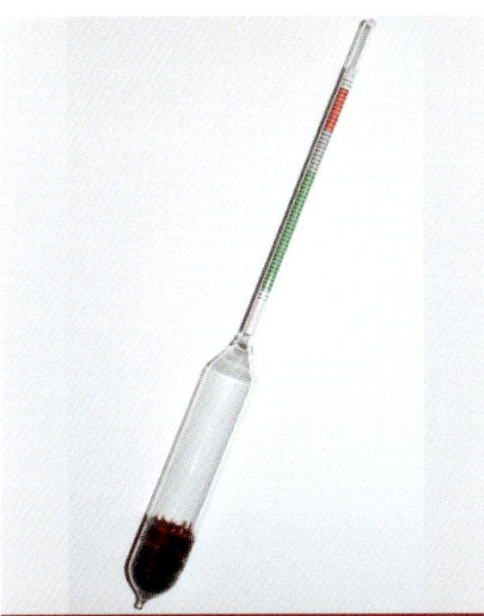
Fig. 7: The colostrometer

Fig. 8: Stored colostrum sample with quality, date harvested and cow ID clearly marked

There are two types of colostrometer: those designed for testing colostrum as it is harvested from the cow at 38°C (100°F); and those for testing at room temperature 22°C (72°F) – make sure you know which one you are using by checking the label.

Colostrometers are criticised for being fragile, having an overlap of high and low antibody specific gravity concentrations and for temperature-based variation. It is, therefore, important to test the colostrum at the correct temperature as results are temperature-related. High temperatures underestimate antibody and low temperatures over estimate antibody. However, colostrometers are a cheap and easy way to evaluate colostrum on farm.

Many vets and farmers now use Brix refractometers. Brix refractometry is traditionally used in the wine, sugar, fruit juice and honey industries, and is a measure of sucrose in a sucrose-containing solution. When used in non-sucrose-containing liquids e.g. colostrum, Brix percentage approximates total solids percentage.[9] As with other refractometers the test solution (colostrum) is placed on the glass surface and a scale read through the eyepiece. The quality is indicated by the scale:

Poor **<20**
OK **20–22**
Good **>22**

Brix, is therefore, objective plus it is temperature independent. Evaluation of colostrum can also be carried out at cow side, making it easier to implement management decisions based on test results.[9] Digital refractometers are also now available will several settings including Brix (Figs 9 & 10).

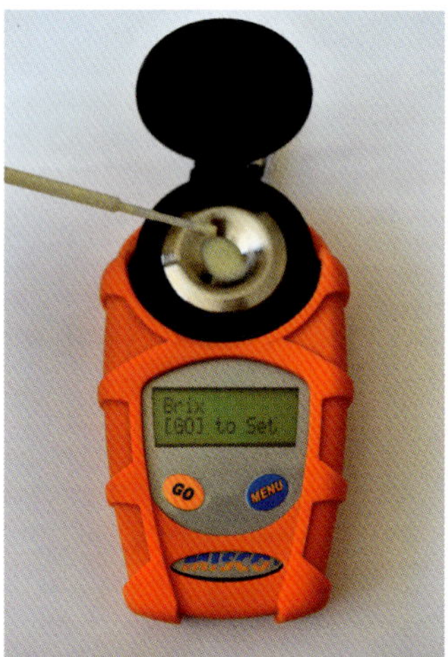

Fig. 9: Digital Brix refractometer, showing sample of colostrum being applied

Fig.10: Digital Brix refractometer, showing a reading for good-quality colostrum

Quantity

Feed 4 litres of high-quality mother's colostrum

The most important factor determining successful passive transfer is the quantity of colostrum ingested or administered to the calf. Dairy calves left with their mother will not ingest an adequate volume of the relatively dilute dairy cow colostrum to meet their antibody requirements.[5] Because we do not know the concentration of antibodies in the colostrum being fed, it is currently recommended that calves be fed 10–12% of their body weight of colostrum at first feeding.[1] For a dairy calf this is approximately 4 litres (Fig. 11).

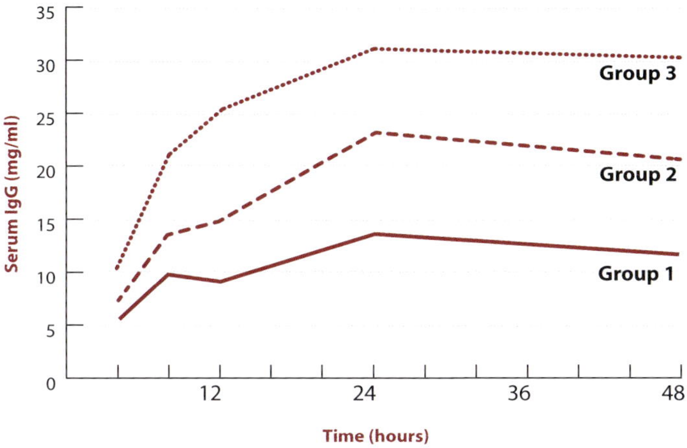

Group 1: 2 litres of colostrum that was low in IgG, fed at 0 and 12 hours

Group 2: 2 litres of colostrum that was high in IgG, fed at 0 and 12 hours

Group 3: 4 litres of colostrum that was high in IgG, fed at 0 hours and 2 litres at 12 hours

NOTE: IgG is the main antibody in colostrum

Fig. 11: Calves fed 4 litres of high quality colostrum have much higher antibody levels than those fed 2 litres of colostrum. (From Morin et al. (1997))[10]

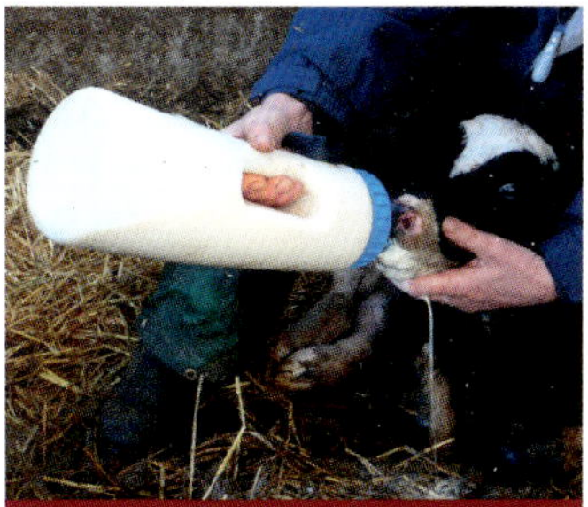

Fig. 12: Feeding colostrum with a nipple bottle

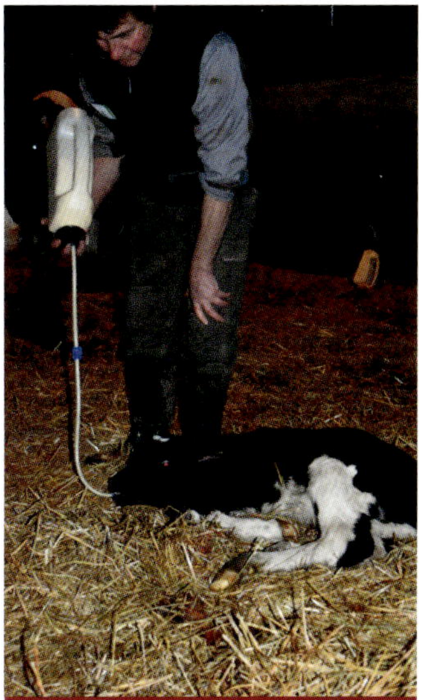

Fig. 13: Feeding colostrum with an oesophageal feeder

Leaving the calf with its mother not only makes it difficult to determine colostrum intake but also increases the risk of infection. Calves left with their mother have a 2.4 times greater risk of FPT in comparison with hand-feeding.[4] Removing the calf as soon as possible from the dam, and feeding 4 litres of mother's colostrum with either a nipple bottle or oesophageal tube (Figs 12 & 13) means you can walk away confident the calf has received an appropriate quantity of colostrum.

Studies have shown that force feeding of 4 litres of colostrum soon after birth is a simple and reliable method to achieve concentrations of antibodies in calves that are well above the desired minimal value of 10g/litre.[11]

Comparison of nipple bottle and oesophageal tube administration of equal volume and antibody mass of colostrum demonstrated that calves allowed to suck a nipple bottle had slightly higher average antibody concentrations. However, this difference was neither statistically nor clinically significant.[5] The author usually advises oesophageal tube administration, which is easy to use and quicker (Fig. 14). Additional advice involves providing another feed of 2 litres of mother's colostrum within 12 hours of birth.

1. Get ready

Before you begin ensure you don't use the colostrum oesophageal feeder for sick calves. Ensure you have separate 2 litre electrolyte bags. You will need as many feeders as there could be calves born in one day.

Any damaged feeders should be discarded they will either damage the soft tissues of the oesophagus or worst still could snap off inside the calf – you should be able to rub the feeder tube on the soft tissue beneath your eye without discomfort.

2. Calf position

The calf must be sat up or stood up. Then push the calf's backside into the corner of the pen, ideally have an assistant to help you and straddle the calf.

3. The Tube

Measure the length of tube to the elbow of the calf – this is how far you can put it down. Dip the end of the tube into the colostrum to lubricate it and allow it to slip down easier.

4. Inserting the tube

Open the calf's mouth by applying pressure to the corner of its mouth or by grabbing over the bridge of the nose and applying pressure to the upper palate or gums. Then pass the tube into the left-hand side of the mouth, as the oesophagus usually sits on the left-hand side of the neck.

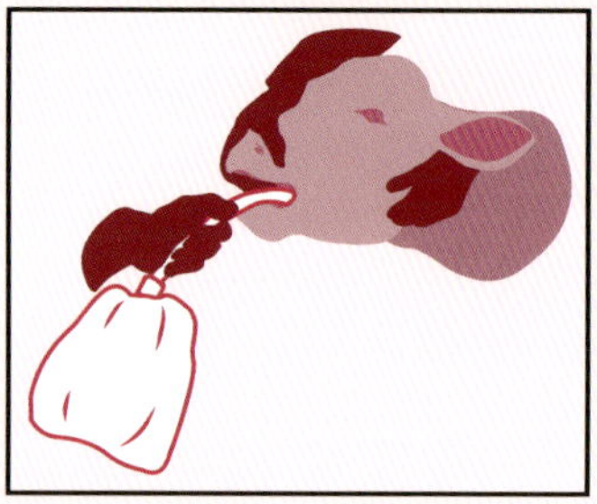

5. Calf head position

When passing the tube ensure the nose of the calf is below the ears. Follow the path of the tongue where there is a ridge at the back. When the tube is over the back of the tongue, the calf starts chewing and swallowing, and so the tube should then go down the oesophagus.

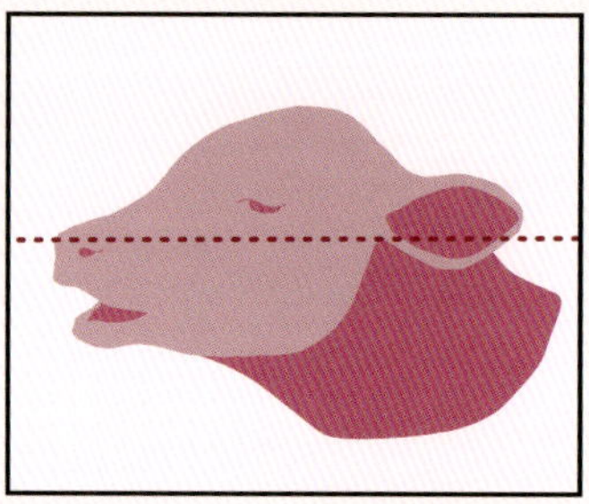

DO NOT pull the calf's head up or trauma will be caused to the back of the calf's throat.

GO SLOW–the slower you go the smoother the passage of the tube. When the calf swallows the tube, passage should be easy

UNCERTAIN if you are in the right place hold the neck of the calf to feel the tube moving down the neck – the ball at the end should be felt quite easily. You cannot feel the tube if it is in the cartilaginous windpipe. If you are still unsure start again–once the tube is in place the calf can change its head position if it likes

6. Adding the fluid

Allow the fluid to pass by gravity. When all the fluid has drained, remove the tube slowly and gently.

6. Clean up

Clean the feeder with warm water, then with 2 litres warm water containing 30 ml bleach – allow this to run through the tube, rinse with warm water again and then hang to drip dry.

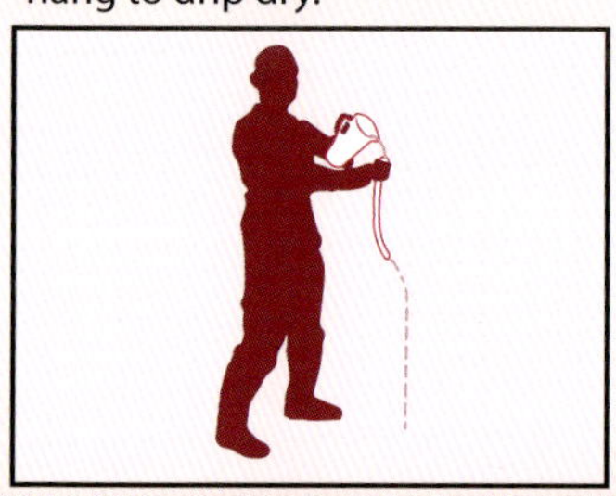

Fig. 14: Use of the oesophageal feeder series – Six-point plan

Quickly

Feed 4 litres of high-quality mother's colostrum within 2 hours of birth and a further 2 litres within 12 hours of birth.

Quality of colostrum is affected by how soon it is harvested. Absorption by the calf is affected by how soon it is fed. In a process referred to as "closure" the efficiency of antibody absorption through the gut lining of the calf decreases with time from birth to being completely closed at approximately 24 hours.[1] When the calf is born cells lining the surface of its gut are able to absorb large molecules, such as antibodies, whole. Absorbed antibodies pass into the bloodstream and are immediately active and able to repel invading infections.[2] The cells lining the gut are rapidly shed so that by 24 hours none is left. The lining of the gut in the newborn calf is like a sliding door or window that is closing rapidly (Fig. 15).

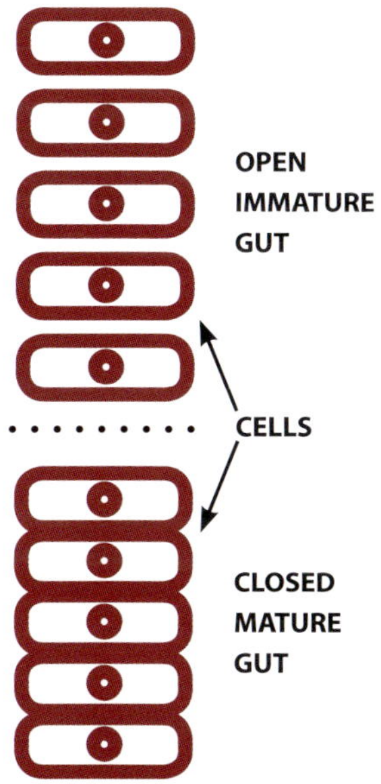

Fig. 15: Gut closure

Despite the length of time to gut closure, antibody transfer across the gut is optimal in the first 4 hours after calving and there is a progressive decline after 6 hours. Calves fed colostrum more than 4 hours after birth are 2.7 times more likely to have FPT than those fed within 4 hours of birth.[4] This is because fewer antibodies are able to get through. Fig. 16 shows how calves fed later have lower antibody levels (NOTE: IgG is the main antibody in colostrum).

Despite the window being closed after 24 hours continued feeding of colostrum over the next several days is still beneficial as it provides local immunity in the gut. This helps to prevent/reduce the severity of bacterial and viral infections that cause diarrhoea in the first few days of life.[13]

The concept of feeding calves as soon as possible is often the most difficult to accomplish. This is often down to lack of staff or time. The benefits of harvesting colostrum early – **First is Gold**, plus gut closure and efficiency of absorption over time, are why this is such an important concept to embrace.

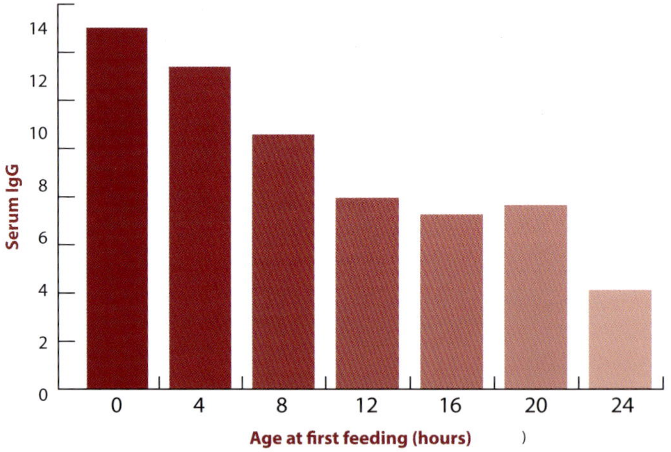

Fig. 16: The earlier a calf is fed colostrum – the higher its antibody levels. (From Stott *et al.* (1979))[12]

Squeaky clean

Feed 4 litres of high-quality mother's colostrum within 2 hours of birth and a further 2 litres within 12 hours of birth. Ensure it has been collected and stored in a hygienic way.

Any feed contaminated with pathogens represents a threat to the calf's health. Colostrum is an important source of nutrients and antibodies, but also represents one of the earliest potential exposures of calves to infections.[1]

Bacteria in colostrum may bind free antibodies in the gut or block uptake and transport of antibodies into the cells lining the gut, thus reducing efficiency of antibody absorption.[14] In addition, prior to "closure", the gut is non-selective in its absorption and so direct uptake of bacteria may lead to septicaemia or diarrhoea (Fig. 17).

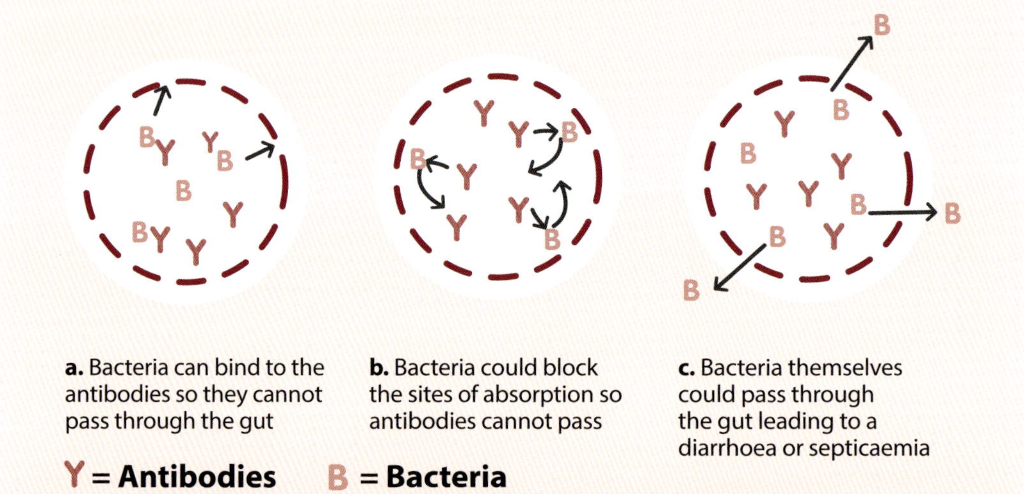

Fig. 17: "Dirty" colostrum can lead to bacteria entering the newborn calf's gut with serious consequences

Just as milk samples from cows can be assessed for bacterial levels, so can colostrum samples. Total bacterial count (TBC) should not exceed 100,000 colony forming units (cfu)/ml and faecal coliforms should be below 10,000 cfu/ml[7] (Fig. 18).

A nationwide evaluation of quality and composition of colostrum on dairy farms in the United States, found that almost 43% of colostrum samples collected had TBC >100,000 cfu/ml and 16.9% of the samples had >1 million cfu/ml.[14] To keep colostrum sQueaky clean involves preventing colostral bacterial contamination plus minimising bacterial growth in stored colostrum. The cow should calve in a clean, straw yard. The udder should be hygienically prepared before collection of colostrum (Fig. 19). Clean technique and milking equipment, which is clean and in good working order, are necessary to minimise bacterial contamination of colostrum[15] (Fig. 20).

TBCs are low or nil in colostrum stripped directly from the udder. This emphasises the need to minimise contamination by milking into a clean sanitised bucket, keeping collected colostrum covered and handling colostrum using clean, sanitised storage or feeding equipment[1] (Figs 21 & 22).

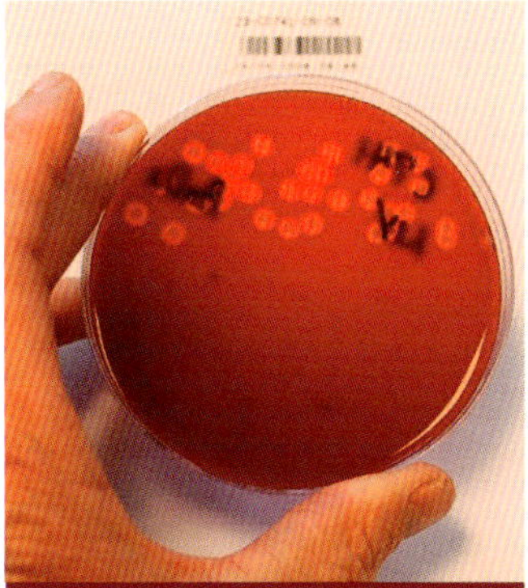

Fig. 18: An agar plate. Colostrum can be sampled in the same way as milk samples and relative numbers of bacteria per ml calculated
(Photo courtesy of M. Millar, School of Veterinary Science, University of Bristol)

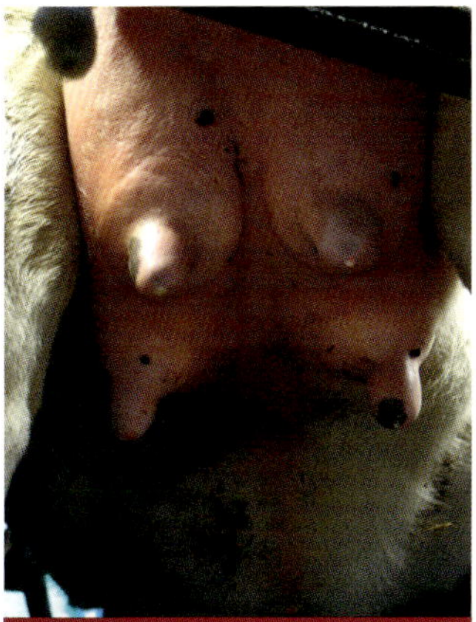

Fig. 19: The udder should be hygienically prepared before harvesting colostrum – exactly the same way as the milking routine

Fig. 20: Clean technique and milking equipment, which is clean and in good working order, are necessary to minimise bacterial contamination of colostrum

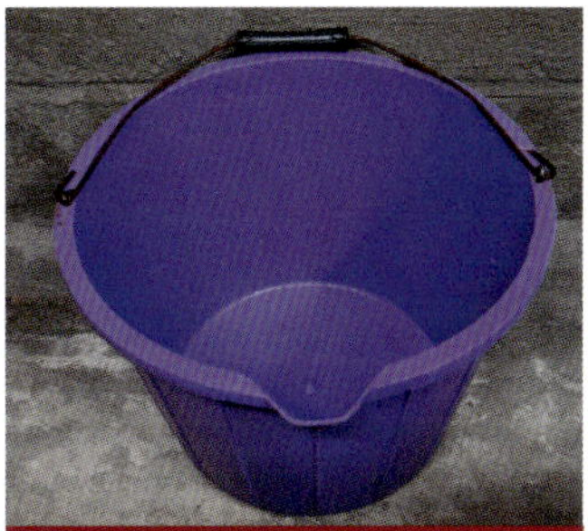

Fig. 21: Colostrum storage equipment must be clean and sanitised before placing colostrum in it

Fig. 22: Colostrum stored in dirty buckets and left at room temperature – this should not be fed to any calves!

Colostrum provides a perfect medium for bacterial growth.[8] Colostrum has nearly twice as many solids as regular whole milk, offering twice the opportunity for bacterial growth: bacterial numbers can double every 20 minutes. Unless it is to be fed right away, colostrum should be frozen or refrigerated within 1 hour after collection[1] (Fig. 23).

Fig. 23: Colostrum stored in the freezer

Fig. 24: Clean 2 litre bottles of ice placed in a bucket of colostrum will drop its temperature rapidly before placing in the fridge

If colostrum is to be refrigerated its temperature must be dropped quickly beforehand. This is because bacteria will continue to multiply for several hours, until the temperature of the colostrum is sufficiently reduced by placing in the refridgerator. Place clean 2 litre bottles of ice within buckets of colostrum to drop its temperature rapidly (Fig. 24). Colostrum can then be stored in a refrigerator for up to 2 days. If it is to be kept longer than this then it should be frozen or a preservative added.

Freeze only colostrum of good quality measured by a colostrometer or Brix refractometer. Frozen colostrum can be stored for up to a year at -18 to -20°C. Store in freezer bags that freeze flat and so will defrost more quickly than milk cartons (Fig. 25). It is a good idea to write cow ID and date of storage in permanent marker on the bag to allow older stored colostrum to be used first, but also if a cow is later identified as having Johne's disease the colostrum can be discarded. Thawing can be done in a clean water bath/bucket no greater than 60°C (140°F) to avoid destroying antibodies. Alternatively, thaw using relatively cheap soup or water heaters with a thermostat that can be set to calf body temperature of 38.5°C (101°F) (Fig. 26). Microwaves should be avoided.

Fig. 25: Colostrum stored flat in 2 litre freezer bags will defrost much more quickly than bulky milk or coke bottles

Fig. 26: Colostrum being defrosted in a soup heater set at 38.5°C

Before feeding the colostrum to the calf, check the temperature – the ideal temperature is calf body temperature, which is 38.5°C (101°F) (Fig. 27). Agitate the colostrum to ensure even heating. If colostrum is too cold, the calf has to use energy to warm it up before it can be absorbed. Newborn calves can also experience cold stress, especially during the winter months. Colostrum provided at body temperature can help warm the calf from the inside out.

Fig. 27: Check colostrum is at the calf's body temperature of 38.5°C (101°F), before feeding

Another simple method to decrease the bacterial load in colostrum is to use potassium sorbate at 0.5% together with refrigeration at 4°C. Potassium sorbate is a preservative (E202) widely used in human food and drink manufacture for its antibacterial and antifungal properties.[13] Potassium sorbate has been to shown to be more effective at inhibiting bacterial growth than refrigeration alone. However, starting with a clean colostrum in the first place is essential – the potassium sorbate only stops growth it does not kill bacteria. Add potassium sorbate to colostrum as soon as possible after it is milked from the cow. Colostrum can be stored in a refrigerator for up to 2 days; adding Potassium sorbate allows it to be refrigerated for 4 days. Add 5 ml of 0.5% potassium sorbate for every 2 litres of colostrum and stir thoroughly.

Fig. 28: A colostrum pasteuriser

An additional tool that is used frequently in the USA and gaining popularity in the UK is pasteurisation (Fig. 28). A temperature of 60°C (140°F) for 60 minutes to batch-pasteurise colostrum is sufficient to maintain antibody activity and colostrum fluid characteristics, while eliminating or significantly reducing important infections.[1] These infections could include *Mycobacterium avium paratuberculosis* (Johne's disease), *Mycobacterium bovis*, *Escherichia coli*, *Salmonella* species and *Mycoplasma* species, which have all been isolated from colostrum.

Pasteurised colostrum can be given to the calf directly or frozen and fed at a later date. There is lots of evidence that pasteurised colostrum improves passive transfer vs fresh colostrum.[16,17] By reducing bacteria numbers in colostrum by pasteurisation there are then fewer bacteria to bind with antibodies to prevent them passing through the gut, and also fewer bacteria to block sites of absorption. This all leads to a better uptake of antibodies by the gut with pasteurised colostrum. Calves receiving pasteurised colostrum have also been demonstrated to have a significantly lower risk for treatment for diarrhoea and treatment for any illness in the pre-weaning period, compared with calves fed fresh colostrum.[17]

The Perfect Udder® bag (Fig. 29), is designed to heat 4 litres of colostrum at a time prior to storage and feeding. This also provides the same efficiency of antibody absorption and serum antibody levels when fed to calves, compared to calves fed colostrum from batch pasteurisers.[18] The Perfect Udder® bag offers the advantage of individually heat-treating 4 litres of colostrum at a time – the volume required by newborn calves.

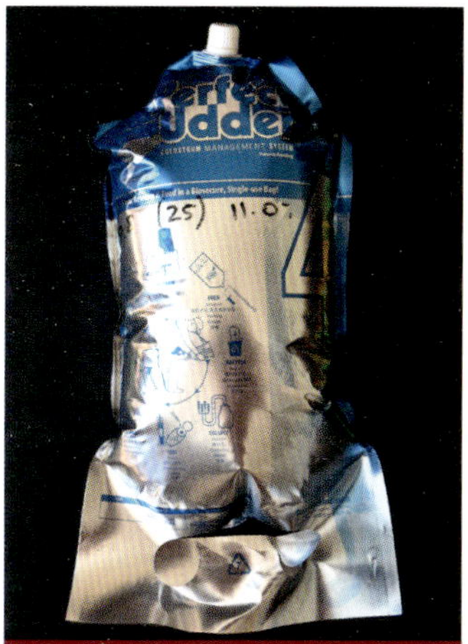

Fig. 29 A Perfect Udder® bag for colostrum. Note: the farmer has written on cow ID on the bag and the date in permanent marker.

The main barrier to using pasteurisers on UK farms is cost. The initial cost is high whether new or secondhand. It must also be remembered a pasteuriser is not a silver bullet and all other components of colostrum management must be adhered to; also, the pasteuriser itself requires good management (see Chapter 3, page 54). However the benefits of pasteurising colostrum to the health and survival of calve's definitely justifies their purchase and use.

If quality colostrum supplies are limited or the cow is infected with Johne's disease, there are a number of colostrum replacers/supplements available. Colostrum replacers must provide > 100 g IgG (the main antibody in colostrum) per complete feeding and nutrients (fat, protein, vitamins, minerals) required by the newborn calf. However, simply examining the mass of IgG provided by the colostrum replacer is not an adequate measure of how well the product works and, ultimately, what levels of antibody will eventually end up in the calf's bloodstream. Colostrum replacers vary by ingredients and quality. This may be due to different manufacturing techniques and quality assurance/testing programmes for different products.[19]

Colostrum supplements usually provide lower levels of IgG and no additional nutrition. They must be fed alongside colostrum. Colostrum supplements typically contain bovine colostrum, cheese whey or egg as sources of antibodies.[15] Products are usually dry powder to mix with water, but beware not to dilute colostrum.[20] Be sure to read and follow the manufacturer's instructions, though, some products are mixed with warm water and fed in an extra feeding, others are added to colostrum, and the number of feedings recommended may vary. Some products packaged in bulk also offer the option of selecting the IgG dose by adding different quantities of powder.[21] As these products are described as "colostrum supplements" rather than colostrum replacers, they are not required to be subject to veterinary medicines licensing.[15]

Colostrum replacers and supplements do have a role in certain circumstances but, usually, if dry cow management and colostrum management is appropriate they are unnecessary. Ultimately, the cow's colostrum is far superior and she has been exposed to infections on the individual farm so will provide the best defences against them.

Quantify

It is no good putting new colostrum management practices in place if we are unable to monitor progress. This focuses on measuring passive transfer. There are several laboratory tests for measuring passive transfer – serum total protein by refractometer is the most reliable test for herd screening.[5] It is a simple, rapid and inexpensive tool. All farm veterinary practices should have the necessary equipment for this (Figs 30–35).

A serum total protein concentration of 5.2 g/dL is equivalent to an IgG concentration of 10 g/L. On a healthy, adequately hydrated calf a serum total protein of 5.2 g/dL or greater is therefore associated with adequate passive transfer.

Regular monitoring of colostrum management by blood sampling calves enables longer-term patterns to be built up (Fig 36). A target of ≥85% of all calves sampled having a result of ≥5.2 g/dL is readily achievable. Monitoring also allows benchmarking between farms (Fig. 37) as well as comparing disease levels with total protein levels (Fig. 38).

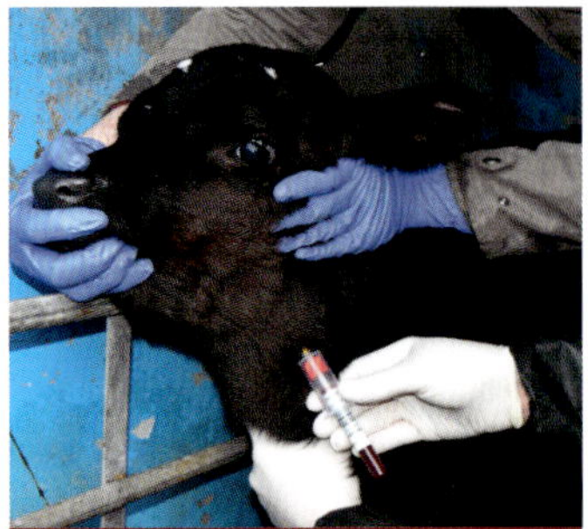

Fig. 30: Calves aged between 1 and 7 days have a blood sample taken

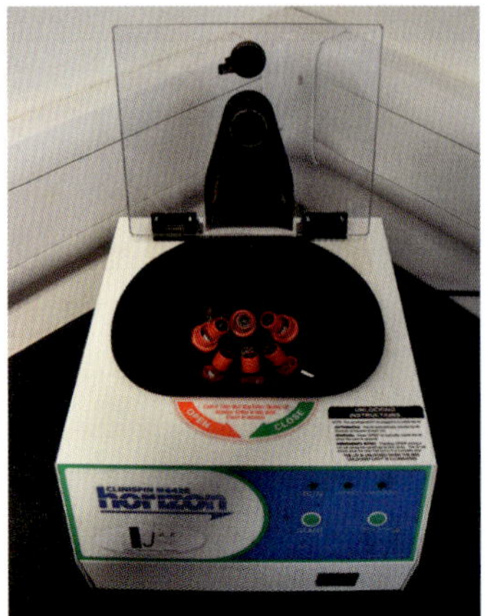

Fig. 31: Blood samples are centrifuged or left to settle overnight

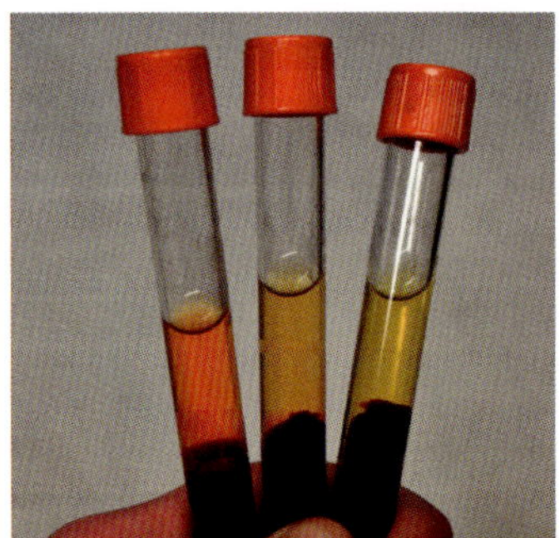

Fig. 32: Serum is the straw-coloured fluid that has settled at the top of the sample

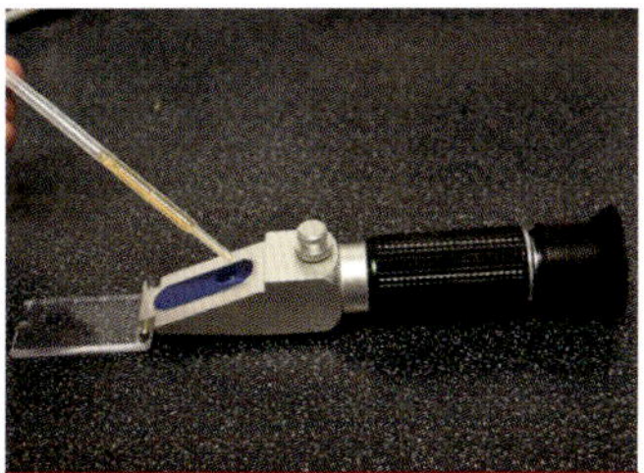

Fig 33: A sample is taken from the serum and placed on the glass surface of a refractometer

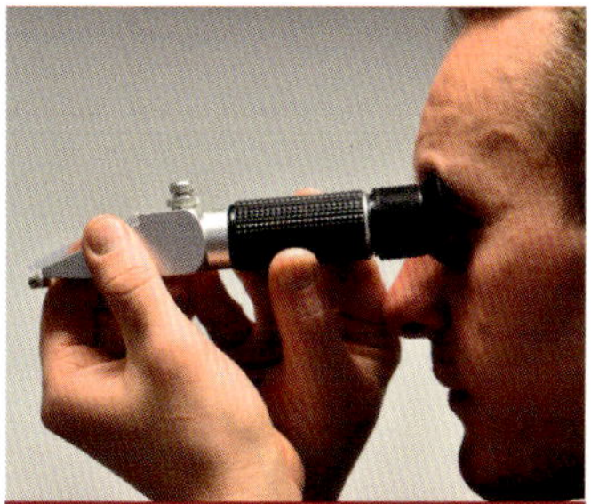

Fig 34: The measurement can be read through the eyepiece

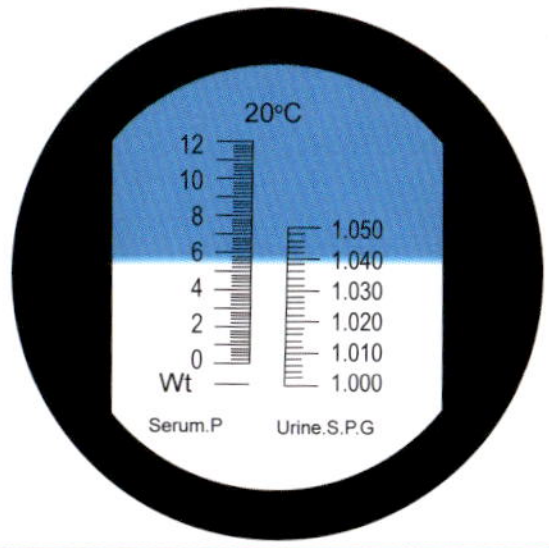

Fig 35: The result is where the blue line intersects the left column

Fig 36: Blood samples are taken from calves every week on this dairy unit. By recording this way trends can be identified as to when colostrum management is good and when it is below standard

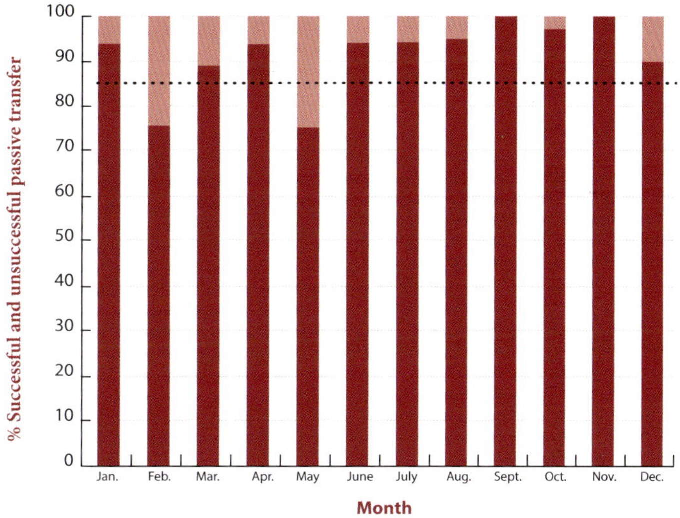

Fig. 37: Benchmarking farms against each other allows a comparison of colostrum management. This graph shows that 5 farms are reaching the target of 85% of calves having adequate colos-trum; 2 farms are just below target and only need to make minor improvements. On the remaining two farms 1 in 3 calves is failing to receive adequate colostrum – changes in management need to be made

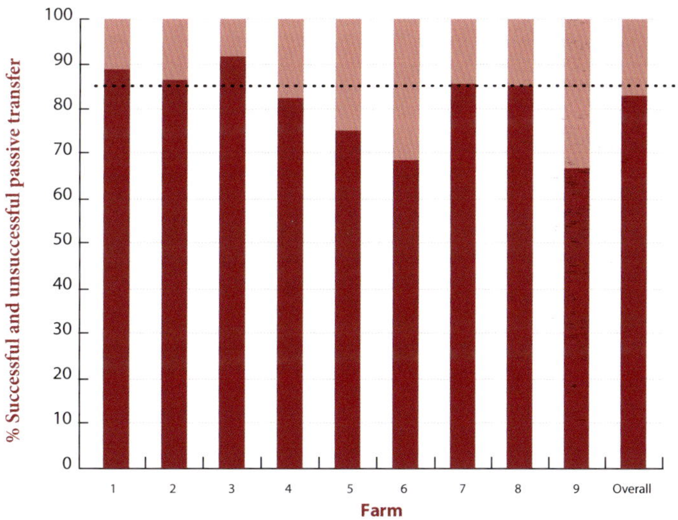

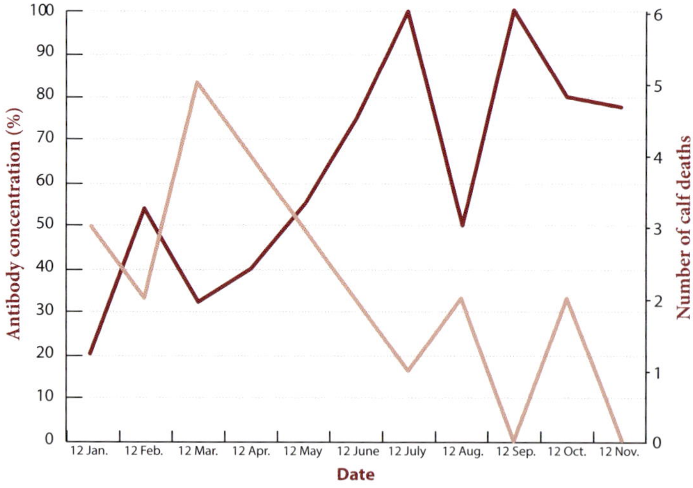

Fig. 38: Monitoring success of colostrum management and calf deaths on this farm showed that as colostrum management improved, the number of calf deaths reduced

% Calves tested with TP ≥5.2 g/dL
Number of calf deaths

A strong association between FPT and the routine measurement of serum total protein in heifer calves has been observed. The likelihood of FPT are nearly 14 times higher on farms that do not routinely monitor total proteins to assess passive transfer compared to those that do.[4] It is logical that improving colostrum management and, ultimately, reducing FPT is a focus of farms that routinely evaluate the passive transfer status of their calves.

Colostrum cleanliness – taking samples and performing TBC's and TCCs (total coliform counts) – can also be performed on a routine basis. Sample colostrum at the point at which it enters the calf, e.g. the oesophageal feeder or nipple feeder into a sterile milk sample pot and refrigerate between 4°C and 8°C before submitting it to a laboratory (Fig. 39). This will allow accurate assessment of how clean colostrum being fed to calves is and adjustments made to collection, storage or administration can be made as necessary.

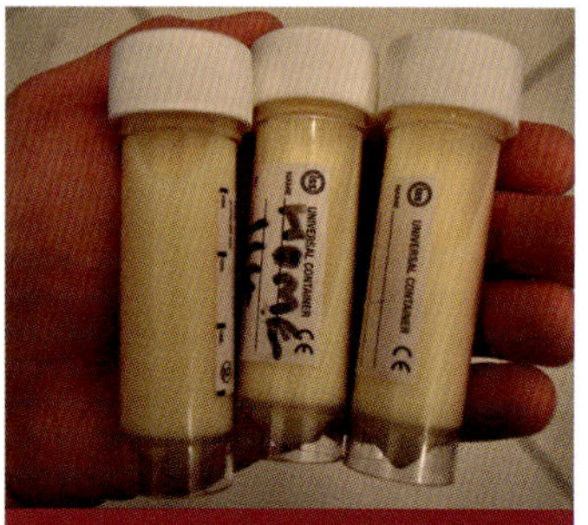

Fig. 39: Samples of colostrum can be taken into sterile pots to measure TBCs and TCCs and so assess cleanliness

Summary

The feeding of an adequate volume of good quality, clean colostrum as soon as possible after birth is arguably the most significant effect a farmer can have on the life of a cow. The importance of providing the best start for future members of the herd cannot be emphasised enough. It is often difficult to focus on this, however, when other jobs such as milking and feeding adult cows take up a significant proportion of the day. Providing simple straightforward colostrum protocols of Quality, Quantity, Quickly and sQueaky clean should enable this focus. They are inexpensive, do not take significant amounts of time and are very effective at reducing disease and death. The ability to then develop programmes to routinely monitor colostrum management and measure success becomes hugely rewarding to all involved.

References

1. Godden, S. (2008) Colostrum Management for Dairy Calves. Veterinary Clinics of North America, *Food Animal Practice* Vol. 24 No. 1.

2. Blowey, R. W. (1999) *A Veterinary Book for Dairy Farmers* (3rd Edition). Farming Press.

3. Davis, C. L. and Drackley, J. K. (1998) *The Development, Nutrition and Management of the Young Calf*. Ames (IA) Iowa State University Press.

4. Beam, A. L., Lombard, J. E., Kopral, C. A., Garber, L. P., Winter, A. L., Hicks, J. A. and Schlater, J. L. (2009) Prevalence of failure of passive transfer of immunity in newborn heifer calves and associated management practices on US dairy operations. *Journal of Dairy Science* 92.

5. Weaver, D. M., Tyler, J. W., VanMetre, D. C., Hostetler, D. E. and Barrington, G. M. (2000) Passive transfer of colostral immunoglobulins in calves. *Journal of Veterinary Internal Medicine* 14.

6. Wells, S. J., Dargatz, D. A. and Ott, S. L. (1996) Factors associated with mortality to 21 days of life in dairy heifers in the United States. *Preventive Veterinary Medicine* Vol. 29 No. 1.

7. Lorenz. I, Mee, J. F., Earley, B. and More, S. J. (2011) Calf health from birth to weaning. 1. General aspects of disease prevention. *Irish Veterinary Journal* 64:10.

8. Potter, T. (2011) Colostrum: getting the right start. *UK Vet* 16 No. 5.

9. Quigley, J. D., Lago, A., Chapman, C., Erickson, P. and Polo, J. (2013) Evaluation of the Brix refractometer to estimate immunoglobulin G concentration in bovine colostrum. *Journal of Dairy Science* 96.

10. Morin, D. E., McCoy, G. C. and Hurley, W. L. (1997) Effects of quality, quantity and timing of colostrum feeding and addition of a dried colostrum supplement on immunoglobulin G1 absorption in Holstein bull calves. *Journal of Dairy Science* 80.

11. Lorenz, I. (2008) Basics of successful dairy calf rearing – focus on immunity and resistance. *Cattle Practice* 16.

12. Stott, G. H., Marx, D. B., Menefee, B. E. and Nightengale, G. T. (1979) Colostral immunoglobulin transfer in calves. III. Amount of absorption. *Journal of Dairy Science* 62.

13. Patel, S., Gibbons, J. and Wathes, D. C. (2014) Ensuring optimal colostrum transfer to newborn dairy calves. *Cattle Practice* 22.

14. Morrill, K. M., Conrad, E., Lago, A., Campbell, J., Quigley, J. And Tyler, H. (2012) Nationwide evaluation of quality and composition of colostrum on dairy farms in the United States. *Journal of Dairy Science* 95.

15. Corke, M. J. (2010) The use of colostrum and colostrum supplements in neonatal calves. *Cattle Practice* 18.

16. Elizondo-Salazar, J. A. and Heinrichs, A. J. (2009) Feeding heat-treated colostrum or untreated colostrum with two different bacterial concentrations to neonatal dairy calves. *Journal of Dairy Science* 92.

17. Godden, S. M., Smolenski, D. J., Donahue, M., Oakes, J. M., Bey, R., Wells, S., Sreevatsan, S., Stabel, J. and Fetrow, J. Heat-treated colostrum and reduced morbidity in preweaned dairy calves: Results of a randomized trial and examination of mechanisms of effectiveness. *Journal of Dairy Science* 95.

18. Kryzer, A. A., Godden, S. M. and Schell, R. (2013) Effect of feeding colostrum that was heat-treated with the Perfect Udder® system on passive transfer of immunoglobulin G in neonatal Jersey calves. *Proceedings of the 46th Annual Conference of the American Association of Bovine Practitioners*.

19. Godden, S. M., Haines, D. M. and Hagman, D. (2008) Improving passive transfer of immunoglobulins in calves. I: Dose effect of feeding a commercial colostrum replacer. *Journal of Dairy Science* 92.

20. Charlton, S. J. (2009) *Calf Rearing Guide*. Context Publishing.

21. Jones, C. and Heinrichs, J. (2011) Colostrum Supplements and Replacer. *Penn State Extension Paper*.

Feeding from birth to weaning

Veterinary Guide To Rearing Dairy Heifers

3

Feeding from birth to weaning

By ensuring we have good colostrum management the dairy heifer has been given the best start to her life – now it is time to capitalise on that initial good work. Nutrition and management practices applied to today's dairy replacement heifers will influence the performance and economic returns of the dairy herd in 2 years time.

The average cost of rearing calves from birth to weaning in the UK is £195.191. The contributions of the component costs are summarised in Figure 1. Feed (colostrum, milk, starter and forage) is the main component at 48.5%, with milk accounting for 37.3% of this.[1]

The objectives of raising the newborn calf to weaning age are optimising growth, minimising health problems and controlling costs. This requires an understanding of the calf's digestive system and nutrient needs as well as the feed and management systems available to meet those needs.

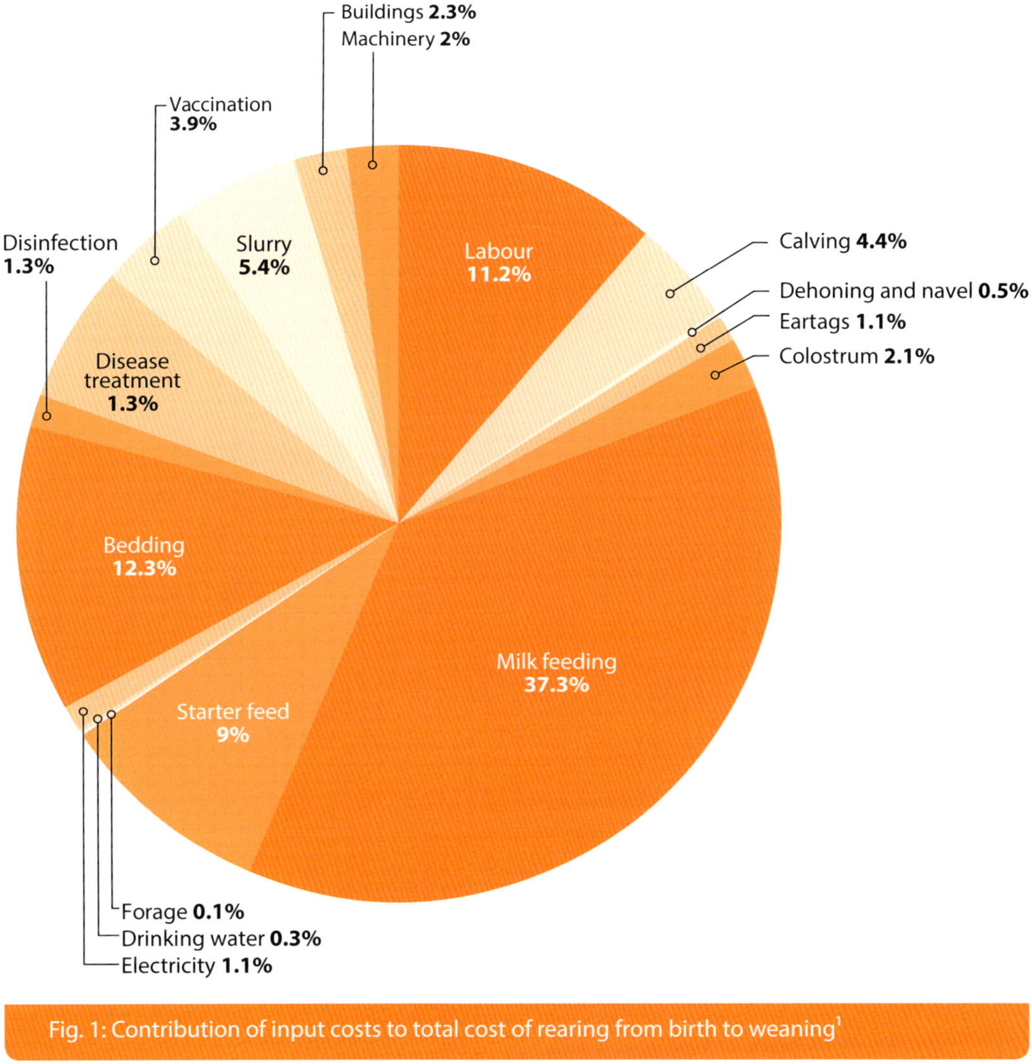

Fig. 1: Contribution of input costs to total cost of rearing from birth to weaning[1]

Feeding from birth to weaning

The digestive system

The calf is a ruminant and like other ruminants (e.g. sheep, goats, camels) has four stomachs – the rumen, reticulum, omasum and abomasum.[2] At birth, the calf's digestive system is underdeveloped. From birth to about 3 weeks of age, the calf is a monogastric, or simple-stomached, animal. The abomasum is the only stomach compartment actively involved in digestion, and milk or milk replacer provides nutrients. As the calf begins to eat dry feeds, particularly grains containing readily fermentable carbohydrates, the rumen takes on a more important role. The stomach compartments grow and change as the calf develops into a ruminant animal (Table. 1).[3]

STAGE	AGE	PRIMARY DIGESTIVE ORGAN	WHAT DIET IS SUITABLE?
Pre-ruminant	0–3 weeks	Abomasum	Liquid (colostrum, milk, milk replacer)
Developing ruminant	4–8 weeks	Abomasum + Rumen	Liquid + solid (starter)
Ruminant	8 weeks +	Rumen	Solid (starter, forage)

Table. 1: The stages of digestion development[4]

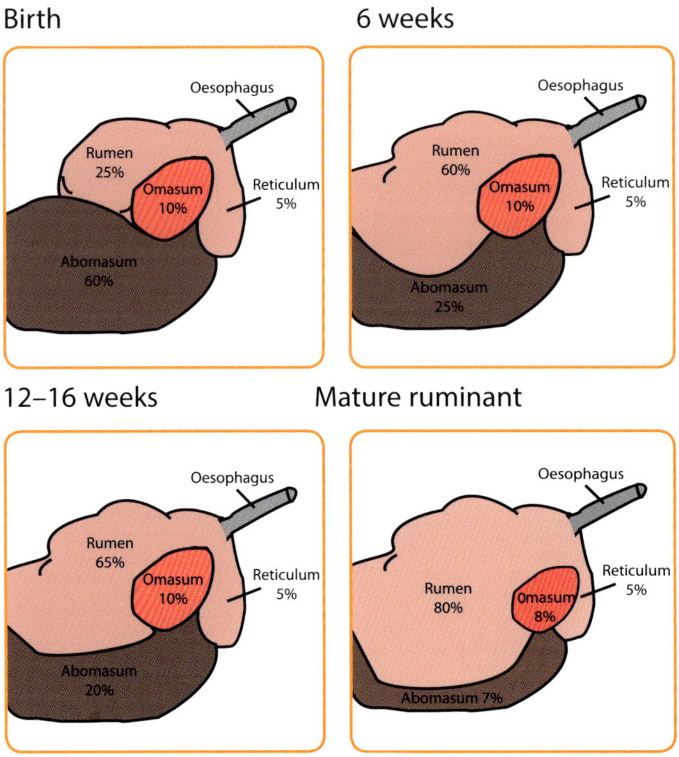

Fig 2: Stages of development of the calf's stomach compartments. (From Charlton (2009)) [4]

Fig. 2 shows how the capacity changes for each of the four stomachs with time. The abomasum continues to function as it did at birth, and it grows in size. However, the reticulum and rumen grow in size and in function; eventually becoming the most important parts of the stomach system. As the stomach develops more fully, the calf begins functioning as a mature ruminant.[3]

Pre-ruminant digestion

During the pre-ruminant phase the solids in milk or milk replacer are digested by the animal's enzymes in the abomasum or small intestine. Milk has to flow directly from the oesophagus into the abomasum, by-passing the rumen and reticulum. This is done by a self-closing channel in the roof of the rumen, known as the oesophageal groove (Figs 3 & 4). When the groove is in the open position, milk passing from the oesophagus will fall into the rumen, become sour and cause a digestive upset. When the groove closes, a "pipe" is formed that transports milk directly through to the omasum and into the abomasum.[2]

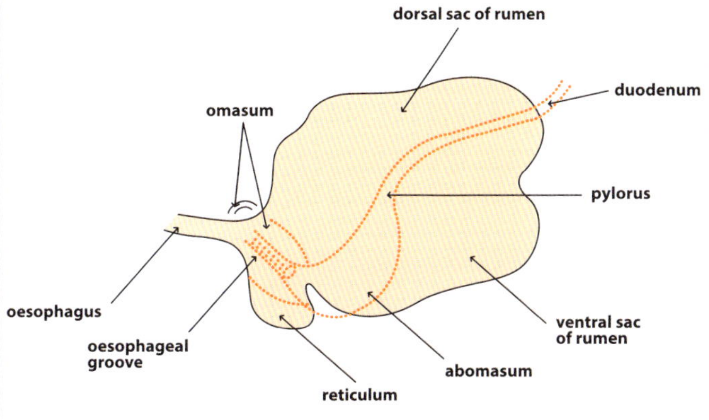

Fig. 3: Position of the oesophageal groove in a ruminating calf (From Blowey[2])

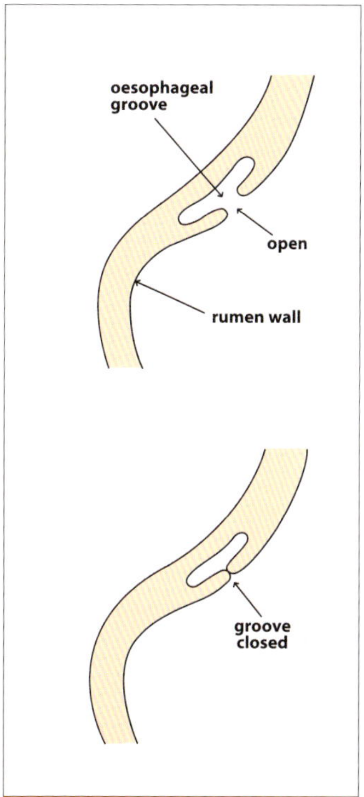

Fig. 4: When the groove closes in the roof of the rumen wall milk flows across the rumen and into the abomasum (From Blowey (1999)) [2]

When the calf suckles there is a reflex action, activated by the bicarbonate in its saliva, which results in the muscular closure of the oesophageal groove. As it gets older, simply the thought of feeding and the sight, sounds and other stimuli associated with the arrival of milk will be sufficient to provoke closure.[2] Oesophageal groove closure occurs whether drinking from a nipple/teat or pail/bucket. Head position affects efficiency of closure,[4] so should preferably mimic the position the calf takes suckling from it mother (Fig. 5). Closure must occur before feeding, and so establishing a feeding routine is vital:[2]

- *Feed at the same time each day, so that the calf knows when to expect its milk*
- *Feed at the same temperature of 38.5°C every time*
- *Feed the same quantities each time, and only make any changes gradually*
- *Let the calves know they are going to be fed. Ideally they should be able to see and hear the milk being prepared for them, so that their anticipation slowly increases*[2]

Fig. 5 Head position can affect efficiency of oesophageal groove closure

Poor groove closure allows milk to fall into the rumen, rather than passing into the abomasum for proper digestion. As there are no digestive enzymes in the rumen, once the milk enters it turns sour leading to a range of clinical signs:

- *Bloat: due to the gas produced by fermenting milk* (Fig. 6)
- *Colic: caused by the pain associated with an inflamed and bloated rumen*
- *Scour: as fermented waste products of milk pass from the rumen into the intestine they often cause chronic diarrhoea*
- *Poor rumen development: this may cause poor growth pre-weaning and often bloat and scour problems after weaning*[2]

Fig. 6: A bloated calf as a result of oesophageal groove closure problems

When whole milk, colostrum or milk replacer with a high proportion of skim enters the abomasum, it forms a curd within 10 minutes. Curd formation is the result of coagulation of milk protein (casein) under the action of two enzymes, rennin and pepsin, and also hydrochloric acid. Milk fat, water and minerals are also trapped in the curd that is retained in the abomasum for digestion.[4]

The fraction of milk that does not form a curd is called whey. Whey is composed of water, minerals, lactose and other proteins (including antibodies). Whey passes directly into the small intestine for absorption and/or digestion within 10 minutes after feeding.[3] Digestion of carbohydrates by the newborn calf is relatively poor, but the exception is lactose or milk sugar. Lactose is digested quickly, by the enzyme lactase, forming glucose and galactose. They are absorbed quickly and provide immediate energy to the calf.[4] Protein is split into amino acids, which can also be absorbed (Fig. 7).

If abomasal milk clot formation is poor, then whole milk passes into the small intestine, where casein cannot be digested. This provides an excellent medium for bacterial fermentation, and scouring results. Poor clot formation can occur because of:

- *Overfeeding, i.e. excessive amounts of milk at each feed*
- *Irregular feeding times*
- *Nervous or stressed calves*
- *Milk fed at the wrong temperature*
- *Skim milk replacer at the incorrect concentrations*
- *Inflammation of the abomasum*[2]

More severe consequences of poor abomasal milk clot formation include abomasal bloat, requiring veterinary attention, or an abomasal ulcer, which could perforate leading to peritonitis, collapse and death.[2]

Milk replacer that does not contain skim milk powder, or only a very low level, will not form a curd in the abomasum. Its digestion efficiency is controlled by the rate of passage.[4]

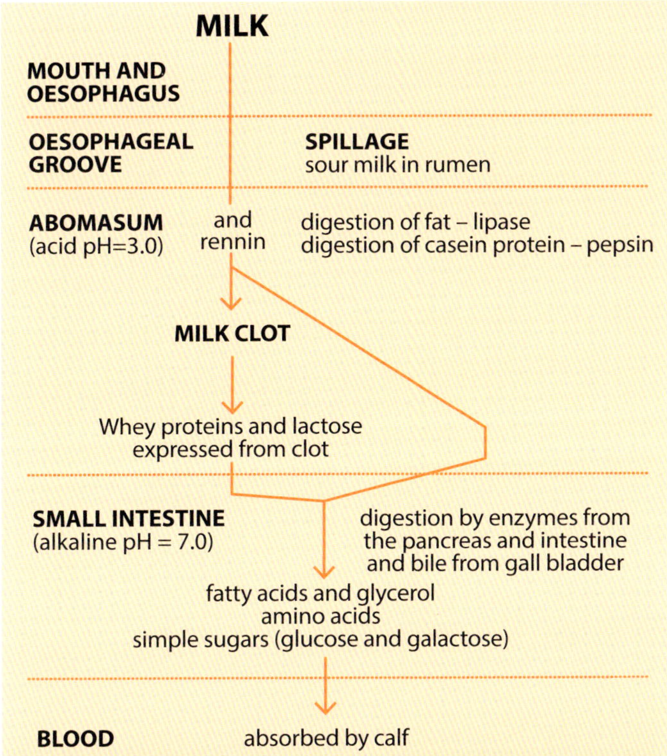

Fig. 7: A "flow" diagram of digestion in the calf
(From Blowey (1999)[2])

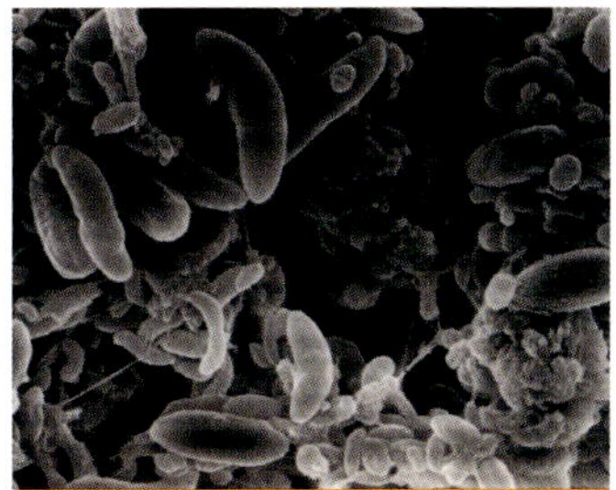

Fig. 8: A healthy rumen population of bacteria, protozoa and fungi

Fig. 9: If a calf eats its bedding the bacteria ingested can contribute to the rumen microflora population

Developing ruminant

The functional rumen acts as a fermentation vat, where microorganisms digest complex carbohydrates and high fibre feedstuffs.[3] The rumen must develop and become fully functional during the period from birth to weaning to have minimal impact on health and growth once milk feeding has stopped. The transition from a "monogastric" or simple-stomached animal to ruminant digestion is a slow and delicate process.[5] Rumination itself is the process of balls of food being regurgitated into the mouth for further chewing and mixing with saliva.[4]

Within a few days of birth, the calf's rumen begins to develop a population of microorganisms. A healthy functional rumen contains bacteria, protozoa and fungi to digest and breakdown complex carbohydrates and high-fibre feeds (Fig. 8). The number and types of microorganisms present are a result of what the calf eats. The oesophageal groove does not function when the calf eats dry feeds; they enter the rumen, where they must be digested by microorganisms or chewed further by rumination.[3]

Rumen bacteria also provide microbial protein (washed out of the rumen). Microbial protein is easily digested and absorbed in the small intestine and contains a good amino acid profile to meet the needs of the growing calf. The rumen microorganisms are able to synthesise nutrients such as B-vitamins and essential amino acids. Digestion of the microorganisms makes these nutrients available to the animal.[4]

In addition to feeds, the environment, bedding (Fig. 9) and hair provide microorganisms for the calf's rumen. The types of rumen microorganisms that multiply are those that best digest and utilise the feeds eaten by the calf.[3] Poor quality colostrum and milk may contain high bacteria levels, including infectious organisms that will begin to contribute to the gut microflora population during development. It is essential to limit this happening via feed, feeding equipment and the environment, as these may have detrimental effects on the colonisation of the rumen by normal "healthy" microorganisms.[5]

Rumen microbes require water to grow properly and ferment feedstuffs. Without water, rumen microbial growth is limited. Water consumed as plain water enters the rumen and becomes available for the microbes' use. However, water consumed in other feeds, including milk or milk replacer, is not available to rumen microbes because it enters the abomasum.[3]

As the calf begins to consume starter concentrates, the microbial population ferments carbohydrates to volatile fatty acids (VFA). Butyric acid, and to a lesser extent propionic acid, stimulates the thickening of the rumen wall, and the development of its characteristic surface papillae[6] (Fig. 10). Papillae are the tiny, finger-like projections that line the inside of the rumen. They are responsible for absorbing nutrients from their surface area and transferring these nutrients into the bloodsteam for energy. If a calf does not have well-developed papillae, it will not be able to capture all the nutrients made available to it through the diet.

Feeding management can drastically affect rumen development. A calf fed milk and good quality hay from 3 days of age will not develop papillae at all, and the rumen wall will be quite thin. This is because the end products of hay digestion include more acetic acid, which the rumen wall does not use for papillae growth and development. Calves with access to large amounts of roughage will have a considerable increase in rumen size. However, this is due largely to the stretching of the rumen tissue, not real growth.[3] A calf fed grain in addition to milk will show a great deal more papillae development and a much thicker, darker and more vascularised rumen wall – this is what we want.

The process of rumen papillae growth is self-generating and allows grain-fed calves to have a tremendous amount of rumen development at an early age – 3-4 weeks. A grain-fed calf can have a rumen with a surface area 40 times greater than that of a hay-fed calf. Feeding grain early allows early rumen development. Calves started on grain late or those that consume too little grain at a young age are at a disadvantage.[3]

Fig. 10: Rumen papillae

Ruminant

By 2–4 months of age the rumen of the calf should be fully functional. The major part of all carbohydrates and proteins are digested by microorganisms in the rumen. The ruminant animal is able to digest plant-based products, including cellulose and hemi-cellulose, by rumen fermentation to yield VFAs and microbial protein. The VFAs are absorbed through the rumen wall and contribute 66–75% of the energy derived from the feed. Carbohydrates (e.g. sugar and starch), microbial protein and undegraded protein which escape rumen digestion, are digested in the abomasum and the end products absorbed through the small intestine wall.[4]

Nutrient requirements

Early nutrition needs to focus on providing adequate energy and protein, while ensuring all required vitamins and minerals are consumed in appropriate amounts and ratios to overall energy intake.[6] The most essential nutrient, water, must be a focus of all feeding programmes.

Energy and protein

Calves require nutrients for maintenance and growth. Maintenance functions are those needed to keep the animal alive; plus maintenance of body temperature in cold and hot climates, immune responses to infectious challenges, and dealing with stressful situations for example transport and poor environment. Growth is the accumulation of new body tissue, which in young calves, before weaning, occurs mainly in the skeleton and muscles.[6] Rates of growth are highest at birth and decline steadily thereafter.

- *A 45 kg calf requires about 325 g of milk solids, or 2.5 Litres of whole milk, for maintenance*
- *A 45 kg calf requires about 380g of milk replacer, or 3.0 Litres for maintenance*[7]

Amounts of milk solids above maintenance can be used for growth.

For calves to grow faster, they need to be fed more milk or milk replacer or, in older calves, eat more starter.

- *Energy is most often the limiting factor to growth*
- *Protein requirements increase as rate of gain increases*
- *The diet must provide the correct balance of energy and protein to support the growth level required*

For example, feeding twice as much of a milk replacer with 20% crude protein (CP) does not provide enough protein for lean tissue growth and the surplus energy is converted to fat.[6] Conversely, feeding a high-protein milk replacer with 26% CP designed for "accelerated growth" at conventional feeding rates (500–600g/day) provides excessive protein to the calves, which is not able to be used for additional growth, because energy is limiting.[8] In this case the excess protein is degraded and the nitrogen excreted in urine.

The thermoneutral zone (TNZ) is the range of temperature within which a calf uses no additional energy to maintain its body temperature. For a newborn calf the range is 15–25°C, for a calf over three weeks of age the range is 6–25°C, and for a growing heifer 0–23°C. Above or below this range, calves must expend more energy to maintain body temperature; in hotter temperatures, they pant and sweat, in colder temperatures they shiver. This increase in energy expended becomes part of the maintenance energy requirement.

If calves are fed the same amount of whole milk or milk replacer as in thermoneutral conditions, less energy is available to fuel growth.

As temperatures become lower then more whole milk or milk replacer is required for maintenance. Table 2 shows the additional milk replacer or whole milk required to maintain growth rates at different temperatures.

Estimates, based on data for older growing cattle[7], indicate increased maintenance requirements of 20–30% during heat stress. Good water access and shade are critical to maintain body temperature in young calves. Transport also increases the required amounts for maintenance. Feed an extra 100 g of milk replacer for calves weighing 43–50 kg, daily for 14 days following transport (in addition to any needed for temperature allowance).[11]

The effects of nutrition on the immune system are often ignored. Calves have an amazing ability to fight disease if the immune system has proper fuel. If it doesn't then that immune system becomes a luxury! Provision of adequate energy and protein is about maintenance and growth, which fuels the immune system too.

Minerals and vitamins

Whole milk, milk replacers and starters normally supply all the minerals and vitamins required by the calf in adequate amounts.

ENVIRONMENTAL TEMPERATURE °C	g/DAY OF ADDITIONAL MILK REPLACER*		LITRES/DAY OF ADDITIONAL WHOLE MILK+	
	Birth to 3 weeks	3 weeks - weaning	Birth to 3 weeks	3 weeks - weaning
20	0	0	0	0
15	0	0	0	0
10	50	0	0.33	0
5	100	50	0.67	0.33
0	150	100	1.00	0.67
– 5	200	150	1.33	1.01
–10	250	200	1.67	1.33

* Based on a basic diet of 6 litres (900 g) of milk replacer with 18% fat and 22% protein containing 18.5 MJ/kg of ME mixed at a rate of 150 g made up to 1 litre with water

+Based on a basic diet of 6 litres of whole milk containing 4.03% fat and 3.28% protein, 22.3 MJ/kg ME on a DM basis

Table. 2: Additional milk replacer or whole milk required, regardless of body weight, to achieve optimal growth with a target growth rate of 0.75 kg/day[10]

Feeding from birth to weaning

Water

Despite being the most important nutrient, water nutrition is often the weakest link on farms. Calves must be supplied with additional water beyond what is provided by the liquid diet. Water has a huge number of roles within the body but with regards to feeding it is important because:

- *Starter intake depends on water intake and both are needed for rumen development*
- *Calves with limited access to water may show up to 38% reduction in lean tissue weight gain*
- *Calves that scour will increase their water intake voluntarily thus improving their chance of recovery.*

Table 3 shows the water requirements for different aged calves.

AGE IN MONTHS	LITRES / DAY
1	5–8
2	6–10
3	8–11
4	12–14

Table. 3: Daily water requirements for different ages of calf

Liquid feed

If left on its mother (Fig. 11), a 45 kg calf will:

- *Nurse 6 to 10 times a day*
- *Consume between 16% and 24% of its bodyweight per day as milk*
- *Consume 8.6–12.7 litres of milk per day*
- *Consume 0.91–1.36 kg of dry milk solids per day*

Modern milk feeding practices differ considerably from a calf nursing a cow, however, it is worth keeping in mind what the calf is designed to do when it comes to feeding.

Fig. 11: A nursing calf

There are many different systems for milk feeding. Choose a system that works on your farm and provides sufficient nutrients to achieve your goals for calf growth, health and weaning age. Consider the advantages and disadvantages of each feed type and feeding method.

- *Availability – is sufficient feed always going to be available for the number of calves?*
- *Consistency – is the feed composition consistent?*
- *Nutrients – does the feed provide sufficient energy, protein, minerals and vitamins for the calf's needs?*
- *Storage and handling – are there facilities to store the feed in a convenient and hygienic manner?*
- *Disease risk – what is the potential for exposure to pathogens?*
- *Economics – what is the cost of the feed and how much will the equipment cost?*[12]

Options

The most common type of milk fed to heifer calves in the UK is milk replacer (55%), followed by whole milk (18%).[1] There are 5 liquid feed options available for preweaned calves:

1. *Milk replacer – a blended powder based on dried products from milk processing*
2. *Whole milk – saleable milk from the bulk tank*
3. *Colostrum – first milk a cow or heifer produces after calving. Low-quality colostrum can be fed later as a feed source*
4. *Transition milk – milk from the first 6 milkings*
5. *Waste milk – milk that cannot be sold as it comes from cows with mastitis or cows under antibiotic treatment. Not recommended.*

Table 4 compares each of the liquid feed options on a dry matter basis

NUTRIENT	COLOSTRUM	TRANSITION MILK (3RD MILKING)	WHOLE MILK	MILK REPLACER (CONVENTIONAL)	MILK REPLACER (ENHANCED)
Dry matter %	24	14.1	12.4	12.5	15.0
Fat % DM	28	27.6	28.8–32	15–20	15–8
Protein % DM	58	36	26	18-23	26
Lactose % DM	11	31.2	39	48	40
Ash % DM	4	5.2	5.6	8	8
Energy MJ/Kg DM	20.1	-	23.4	17.1–19.2	19–19.5

Table. 4: A comparison of different liquid feed options for preweaned calves (From Charlton (2009)[4]

Economics

When it comes to calculating comparative feed costs, this should be done on a dry matter basis and be consistent with units.

Examples:

1. Milk replacer (typical dry matter is 95%)

   Cost of 1 kg of dry matter
   = (cost per bag ÷ weight per bag) ÷ 0.95

   = (42.50 ÷ 25) ÷ 0.95

   = £1.79

2. Whole milk (typical dry matter is 12.5%)

   Cost of 1 kg of dry matter
   = market price per litre of milk ÷ 0.125

   = 25 ppl (March 2015) ÷ 0.125

   = £2.00

In this example, in March 2015, the average cost of 1 tonne of milk replacer was about £1700 (be aware of a significant range of composition, ingredients and feeding guidelines) and milk price was 25 ppl. If the cost of milk powder increased or milk price dropped much more, then feeding whole milk could be considered. Perform this simple calculation to keep informed.

Collection, storage and feeding

Providing a safe and clean liquid feed for the calf is essential. The udder should be hygienically prepared before collection of colostrum, transition milk and whole milk. Clean technique and milking equipment, which is clean and in good working order, are necessary to minimise bacterial contamination. Prepare milk replacer in clean buckets, using clean equipment. Minimise contamination by storing in clean sanitised buckets (Fig. 12), keeping collected liquid covered, and handle using clean, sanitised storage or feeding equipment (Figs 13 & 14).

Fig. 12: Collect liquid feed into clean sanitised buckets. These buckets have been labelled with a permanent marker pen at 12 litres so it is known how much is stored

Fig. 14: Clean feeding equipment

Fig. 13: Feeding calves through dirty equipment poses a huge risk to their health by exposing them to large numbers of pathogens. It is unacceptable on both health and welfare grounds

Liquid feed provides a perfect medium for bacterial growth – numbers can double every 20 minutes. Milk replacer should be prepared immediately before feeding. Unless it is to be fed right away, colostrum, transition milk and whole milk should be frozen or refrigerated within 1 hour after collection. See Chapter 2, page 26 on keeping colostrum squeaky clean – which can also be applied to transition milk and whole milk.

Have a dedicated area within the calf buildings or nearby for storage and preparation of liquid feed, plus for cleaning and storage of all feeding equipment to make the task of feeding calves easier. It also becomes more efficient, reduces the risk of feed contamination and makes the whole process more enjoyable (Figs 15 & 16).

Fig. 15: Clean, organised, dedicated feed preparation and storage area

Fig. 16: A further example of a clean, organised, dedicated feed preparation and storage areas

What not to feed

- Milk with antibiotic residues – milk from cows receiving antibiotics will expose bacteria in the gut to sub-therapeutic levels, potentially leading to the development of antibiotic resistance in gut bacteria[13]
- Milk with high bacterial counts – cows with mastitis pass high numbers of bacteria in their milk. This is the reason this milk is not considered safe for human consumption. In a young calf the risk of infections in the gut is very high if fed this milk. Calves that suck each other immediately after drinking mastitic milk could inoculate the immature teats and eventually cause heifer mastitis[3]
- Milk from cows with infectious disease, e.g. Johne's disease, *Salmonella*, *Mycoplasma*, and *Leptospirosis*. These diseases can all pass via the milk and so calves should not be fed from these cows. Accurate recording and identification of infected cows is very important to ensure this does not happen (Fig. 17)

Fig. 17: A cow with Johne's disease clearly identified with a red eartag to prevent her milk being used to feed calves

Milk replacer

Milk replacer offers a consistent and convenient feed for calves with less expense than whole milk (Fig. 18). With diseases such as Johne's, *Salmonella*, *E. coli*, *BVD*, *Pasteurella* and *Mycoplasma* all able to be transmitted through unpasteurised milk; a milk replacer offers an alternative with none of these risks.

Fig. 18: Milk Replacer

What is in a milk replacer?

A milk replacer is a blended powder formulation based on dried products from the milk-processing industry (Fig. 19), fats and oils, minerals, vitamins and additives. They are a complete, balanced, nutritional feed for calves from birth to weaning.[4]

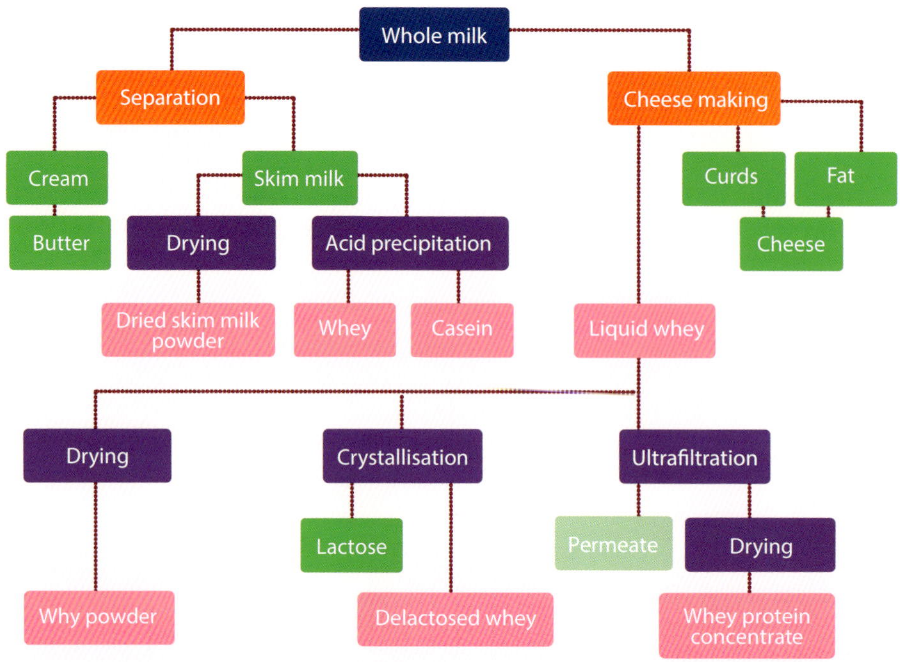

Fig. 19: Whole-milk processing and associated by-products used for calf milk replacers (pink) (From Cooper and Watson[14])

Milk replacer's are either based on skim milk powder (SMP) or whey powder:

- Skim milk powder is a by-product of butter and cream manufacture (the butterfat is "skimmed" off the top of the milk). Skim milk retains all major milk proteins, which are typically around 80% casein and 20% whey, and is capable of forming a functional clot in the abomasum[14] (see Preruminant digestion on p. 37). The price of SMPs has increased since EU subsidies for their use were removed, therefore, they are now lower in their SMP content, typically by around 20–50%. There is no legal definition as to what constitutes the minimum inclusion of skim milk in a "skim milk powder" and so where SMP features low down on the ingredient list view it with caution. At around 5% protein inclusion, casein only forms a weak clot[15] and so powders containing this level of casein or less increase the risk of diarrhoea.[14]
- Whey milk powder is a by-product of cheese manufacture (Fig. 20) made up primarily of lactose, whey proteins and minerals. Typically, there will be 75% lactose and 10–15% whey proteins. Whey-based powders generally use whey proteins as the predominant source of protein.[14]
  Unlike SMPs, whey powders do not form a functional clot in the abomasum. Casein in a SMP can be retained in the abomasum for up to 6 hours after ingestion, whereas whey proteins leave the abomasum within 3 hours[6]. This is not a problem under normal circumstances, as whey proteins are readily digestible in the intestines.
  Whey proteins may be produced by a variety of manufacturing processes, that affect both their feeding quality and digestibility.[14]

Fig. 20: Whey powder is a by-product of cheese manufacture. At this stage in manufacture the whey is separated from the curds

Milk components are the most expensive in milk replacer formulas.[6] Therefore both skim and whey-based milk replacers may contain some non-milk proteins, and will definitely contain some form of blend of plant-based fats to replace milk fat removed during processing.[14]

The digestive enzymes in the young calf are especially adapted to the digestion of milk proteins. Digestion of non-milk proteins is limited and develops gradually over the first few weeks. No non-milk proteins can equal the growth, performance and health resulting from use of milk proteins.[6] To be used successfully in milk replacers, therefore, alternative proteins must be well used by the calf and mix well.

The most common forms of non-milk proteins found in milk replacers are soya-based protein, hydrolysed wheat gluten and pea protein.

- *Soya proteins are an inexpensive form of moderate quality protein. However, soya contains several anti-nutritional factors that may cause issues in calves if used in excess*
- *Hydrolysed wheat gluten is a by-product of grain carbohydrate extraction. It contains a high level of protein, is very digestible and free from anti-nutritional factors. It is the most commonly used form of plant protein in UK milk replacers*
- *Pea protein is a highly digestible, concentrated form of protein that contains no anti-nutritional factors. Pea-protein sediments out of solution rapidly though, limiting its inclusion in milk replacers. It may be worth inspecting buckets at the end of feeding to check if residual solids remain if pea protein appears on the ingredients list*[14]

In the EU, feeding animal-based fat sources to ruminants is not permitted, with the exception of some marine-based oils. Extra fat sources must be added in both skim and whey powders to replace butterfat removed during processing. These are invariably emulsified vegetable fats, most commonly coconut, palm and soya. Ideally a blend of fat sources is used, with the aim to replicate those found in milk fat.[14]

Care should be taken to mix a milk replacer at the water temperature specified by the manufacturer and to not mix too vigorously, otherwise fat suspensions break down. Also scouring and decreased performance can result when fat is not incorporated properly into the milk replacer.[6]

Feeding from birth to weaning

The feed industry is full of various additives that are reported to improve calf health and growth. It is unlikely any magic bullets truly exist though! Several studies have shown pre-ruminant dairy calves respond positively to oral antibiotics in milk powder. They are not a substitute for good management, however. The increasing risk of antibiotic-resistant strains of bacteria developing mean this practice is one of last resort and can only be justified when diagnostic testing has proved the need for incorporating antibiotics into a milk powder. Currently, the only available antibiotic for inclusion in milk powders in the UK is Chlortetracycline (CTC) oral powder.

The VMD (Veterinary Medicines Directorate) states CTC oral powders:

- *Are only authorised for the treatment of individual animals, following a clinical assessment of those animals by a veterinarian; and*
- *Are authorised to be administered for 7 days*

In addition:

- *Veterinarians must be able to justify their prescriptions for CTC oral powder and keep appropriate records*
- *Farmers should seek advice from their veterinarian, and must not request prescriptions that are against the veterinarian's professional judgement*

There will always be several management strategies that can be employed rather than having to use antibiotics in a milk powder. It is essentially "papering over the cracks". It is also unlikely the practice will remain legal, in the future.

Probiotics are live bacterial cultures designed to increase colonisation of the gut with species that produce favourable effects and compete with pathogens for nutrient sources and attachment sites.[6] Results of research trials have been variable.[3] Prebiotics are compounds that are indigestible by the calf's digestive enzymes but are usable by rumen microorganisms. They provide a food for growth of beneficial bacteria in the gut, or compete with disease causing bacteria for attachment sites on the gut lining.[6] The jury is out on both of these products; a focus on good nutrition should mean there is no need to look for "bonus additives".

Choosing a milk replacer

The type of milk replacer best suited for a farming operation will depend largely on the growth requirements of the animals. It should also be suitable for the chosen method of feeding (e.g. restricted, ad-lib or accelerated).[14] (See Liquid Feeding System on p. 59.)

Milk replacers manufactured by reputable companies are produced under strict quality-control guidelines and are of a high quality. Manufacture of uniform, high-quality milk replacers is a highly specialised process requiring special attention to both ingredient-quality and process-quality control. Reputable manufacturers use only high-quality ingredients that are monitored closely for microbiological contamination, heat damage, oxidative damage and other quality measures.[16] Like all things in life you get what you pay for – calf milk replacer represents an investment in the future of the dairy herd – so invest in the best you can afford.

A basic assessment can give information on the quality of a powder by examining it both dry and when reconstituted. The following guidelines have been adapted from the Bovine Alliance on Management and Nutrition.[17]

Dry Powder (Fig. 21)

- Colour: Cream to light tan, free of lumps and foreign material. If powder is orange to orange – brown in colour and has a burned or caramelised smell, the product may have undergone Maillard Browning (non-enzymatic browning) as a result of excessive heat during storage. If the product has "browned", there will be some loss of nutrient quality and product palatability
- Odour: Powder should have a bland to pleasant odour. If odour is characterised as smelling like paint, grass, clay or petrol, the fat portion of the product may be rancid

Fig. 21: Dry milk powder

Reconstituted liquid (Fig. 22)

- Mixing: The product should go into solution easily. Milk replacer should be mixed at manufacturer's recommended water temperature. Mix until all powder is in solution or suspension without clumps of undissolved powder on the surface of the solution or at the bottom of the pail. Ingredients that are in suspension but are not soluble will settle out of solution (form a sediment at the bottom of the container) if allowed to stand without agitation. This sediment layer will be more apparent as the fibre content and/or level of added minerals and/or medication increases. In some feeding situations (automatic feeders, nipple bottles, etc.), milk replacers containing significant quantities of insoluble components may not be suitable. Care should be taken not to over mix. If agitation is continued after product is in solution, excessive foaming can occur or the fat portion of the product may separate and form a greasy layer on the surface.

Fig. 22: Reconstituted milk powder

- Colour: Cream to light tan
- Odour: Pleasant with no "off" odours noted
- Flavour: Milky with no "off" flavours. Some milk replacers are supplemented with organic acids. These milk replacers will have a "tangy" (sweet tart) taste. This should not be confused with the "off" lactic acid taste found in sour milk

A visual assessment should be coupled with an assessment of the milk replacer label. This is not straightforward, but it is possible to get a general impression.

Feed ingredients are legally listed in descending order of inclusion. Sources of protein are usually evident, but not relative percentages of each (including what proportion are from vegetable sources), their digestibility or amino acid balance. Equally, individual fat sources are not usually listed at all, nor their relative percentages.[14] Any medication must also be declared with details of use and withdrawal time if applicable.[4]

Little nutritional information is given on a label: crude protein, fat content, crude fibre and ash, plus vitamins A, D and E, copper and occasionally selenium. The declared minerals generally vary little between milk replacers, so do not usually warrant routine inspection.[14]

Crude protein and fat content are self-explanatory. Crude fibre may be used to help identify powders with high plant-protein content. Powders containing >0.2% crude fibre are likely to have considerable plant protein present, although some types of plant protein may contain little fibre (e.g. soy protein isolate) so low crude fibre does not guarantee low plant protein inclusion. Crude fibre is, therefore, best interpreted in conjunction with the ingredient list.[14]

Ash levels represent the overall minerals in the milk replacer – mostly sodium, potassium and chloride. Whole milk from cows contains approximately 6% ash, so a milk replacer with an ash level of 6–7% will be appropriate for a calf. Anything above this should be avoided as it poses specific health risks to the calf – cellular dehydration, reduced dry matter intake, scour and abdominal pain. Furthermore, there is no energy value in ash, i.e. it has no benefits.

Feeding recommendations – mixing and feeding levels, should also appear on the label.

Due to the limited information available, contacting manufacturers of milk replacer may yield further information. It is also possible to have milk replacer analysed independently, which will give a lot more detail on constituents.

Preparation of a milk replacer

Calves like a "boring" life – they want the same things, in the same way, every day. Preparing a milk replacer accurately and consistently for every feed presents the calf with the correct nutrition and will maximise growth and health. Preparation requires focus on:

- Cleanliness – Mixing and feeding equipment must be hygienic to maintain high-quality feed for the calf (Fig. 23)
- Mixing – The technique depends on the feeding system used. It involves both the method of mixing (whisk, computerised feeder, etc.) and mixing time[4]
- Concentration – The amount of dry matter in the liquid milk: 10–15% depending on rearing system
- Temperature
  Milk – At calf body temperature of 38.5°C (101°F)
  Water – The temperature of water used to reconstitute the milk replacer depends on type. Milk replacer that is "agglomerated" during manufacture (fat droplets are encased in protein and lactose), for improved wettability, are best mixed with warm water (40–45°C, 104–105°F). Other milk replacers are best mixed with hot water (60°C, 140°F)[4]. Check the manufacturer's recommendations

Fig. 23: Keep feeding equipment clean

Computer-controlled milk feeding systems will mix the milk replacer with water automatically, although this must be monitored regularly (see Liquid feeding systems on p. 59). For hand-mixing as in a bucket feeding system, see Fig. 24.

Notes of caution:

1. There is a big difference between:

   a) mixing 150 g powder in 1 litre (13% solution – 1.15 litre mixed) and;

   b) 150 g in 850 ml to make up to 1 litre (15% solution)

   A 45 kg calf drinking 6 litre a day of the 13% milk will grow 20% more slowly than its twin drinking 6 litre of the 15% milk (Fig. 25). Weigh the powder and add the appropriate volume of water to make 1 litre for the required concentration.

2. Milk replacer will vary in density between batches and even within a bag, i.e. If stood up the powder is more compacted at the bottom of a bag. Using a pre-weighed scoop and assuming the powder in it will be the same every time can lead to inaccuracies. A level scoop could weigh from 375 g to 450 g; so when you think you are feeding 900 g per day you could be under-feeding by 150 g. This equates to 8.4 kg over 56 days so it is advisable to weigh powder every time.

3. Keep the milk scoop clean, and don't let milk powder build up around the edge. This will reduce the amount of milk powder measured out, and the dried-on powder can eventually become contaminated – leading to a risk of infection to the calf.

Add 1

Put HALF the required volume of warm or hot clean WATER (45–60°C, 110–120°F) into a "clean" container

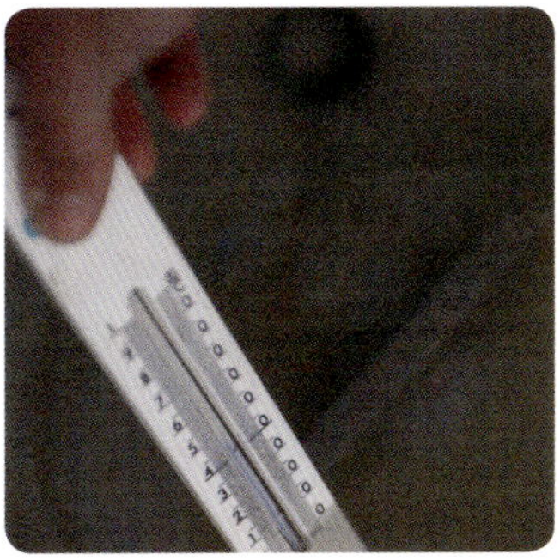

Measure 2

MEASURE the amount of POWDER required and ADD to water by sprinkling on the surface

Mix 3

MIX for 3 minutes using a good wire whisk or a paint mixer, to the bottom of container, until smooth and in suspension

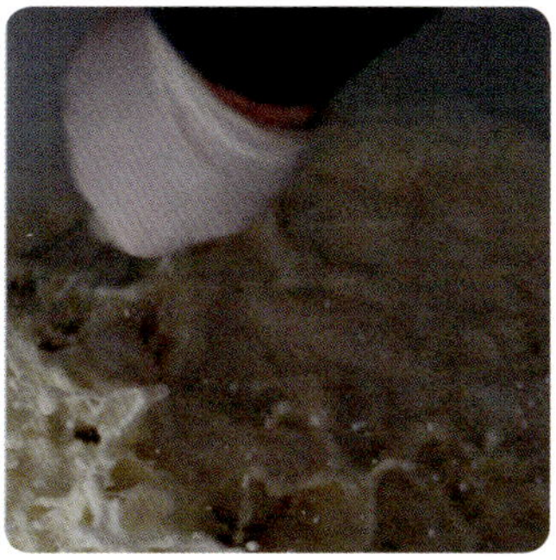

Add 4

ADD COOL WATER to make up to required total litres
MIX for 1 minute

Check 5

CHECK TEMPERATURE of milk solution (38–40°C, 100–104°F) using a thermometer and FEED required amount to calf

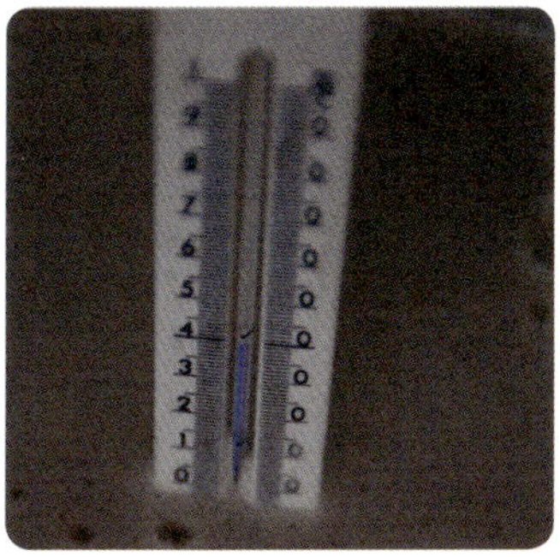

Fig. 24: The Five Steps For Hand Mixing: Add, measure, mix, add, check (From Charlton (2009)[4])

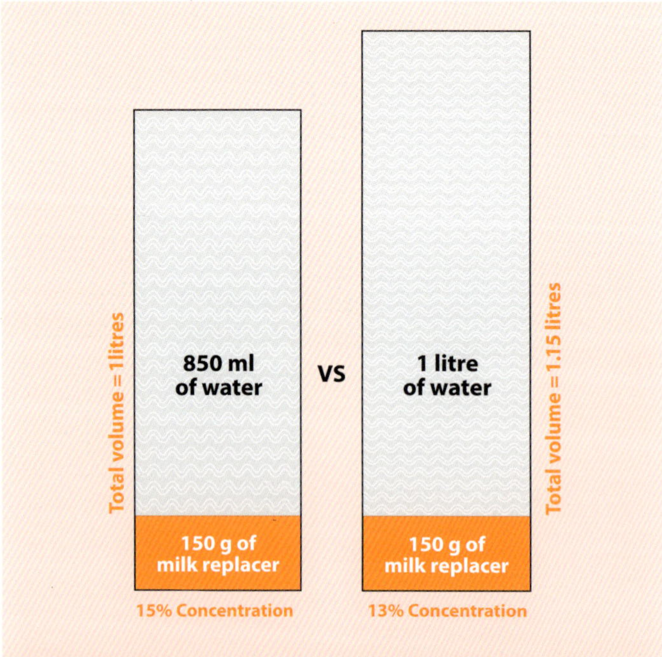

Fig. 25: How mixing milk replacer differently can have different results

Whole milk

Whole milk is an excellent feed for calves – it is nutritionally balanced for their needs. Whole milk is not waste milk or milk from cows treated with antibiotics, which should be discarded and not fed to calves. The purpose of a milk replacer is it has to be cheaper than whole milk otherwise why use it? A milk replacer is also going to be poorer quality. Table 5 compares a 20% crude protein and 20% fat conventional milk replacer, with whole milk.

WHOLE MILK	MILK REPLACER 20:20
12.7% Solids	12.5% Solids
27% Protein	20% Protein
30% Fat	20% Fat

Table. 5: Whole milk vs. milk replacer

Whole milk has 35% more protein and 50% more fat when compared to milk replacer on a dry matter basis. This does not include all the other components that make up these two feeds. It stands to reason, therefore, that calves do very well on whole milk. However, whole milk out of the bulk tank can vary considerably in constituents – consider somatic cell counts and bactoscan, protein and butterfat levels (Fig. 26). Milk replacer – if prepared correctly, provides a very consistent product – with no variation, which calves prefer. To ensure a consistent product with the benefits of whole milk, pasteurisation or acidified milk feeding is required. Both also enable the feeding of waste milk.

Troubleshooting

Milk replacers are often blamed in cases of "nutritional scour" or poor growth of calves. It is worthwhile ensuring that all those involved in feeding calves, including weekend and holiday cover, are aware of feeding protocols, and that discussions take place with everyone. If a problem persists, an investigation, in conjunction with a veterinarian, should take place. Table 6 lists the lines of investigation.

"NUTRITIONAL" SCOUR	
Gut overload	Volume and/or concentration should be increased slowly from birth to the desired endpoint
Bacterial contamination	Mixing buckets and equipment should be cleaned after every feed. Milk feeders or buckets should be cleaned with detergent after every feed
Incorrect mixing temperature	Heat damage at mixing, e.g. powder mixed with hot water from the boiler and topped up with cold. Aim to feed at 38.5°C. A thermometer should be used to assess temperature, not hands. Keep temperature <50°C
Poor mixing	Start with a small amount of water in the bucket, add the milk powder and whisk thoroughly. Top up to the correct amount with warm water, mixing as you go. Serve immediately.
Poor powder digestibility	High inclusion of partially denatured whey or casein in milk replacer; high inclusion plant proteins +/- anti-nutritional factors
POOR GROWTH RATES OR PERFORMANCE	
Rumenal "drinking"	To ensure good oesophageal groove closure, calves should be fed at the same time each day, same height, same system (e.g. buckets or teats), same temperature and with the same product (i.e. do not alternate powder and waste milk)
Insufficient powder	Volume-based scoops need to be calibrated for weight at every feed (even the same powder from the same manufacturer); bulk density can vary considerably
Low powder relative to the animal's needs	An appropriate (calculated) amount of powder needs to be fed for desired growth rates and altered accordingly according to environmental conditions (e.g. low temperatures) and other stresses
Incorrect concentration	To make up a 12.5% /125 g/litre solution, the milk powder needs to be topped up to 1 litre with water, not added to 1 litre of water (which would make a 111g/litre solution)
Lack of supporting feeds	A digestible calf starter should be available for calves to experiment with from day four. As animals grow, nutrient intake from starter forms a considerable proportion of dietary intake.

Table 6: Suggested lines for investigation in situations of suspected "nutritional" scour or suboptimal calf performance, Adapted from Cooper and Watson (2013)[14]

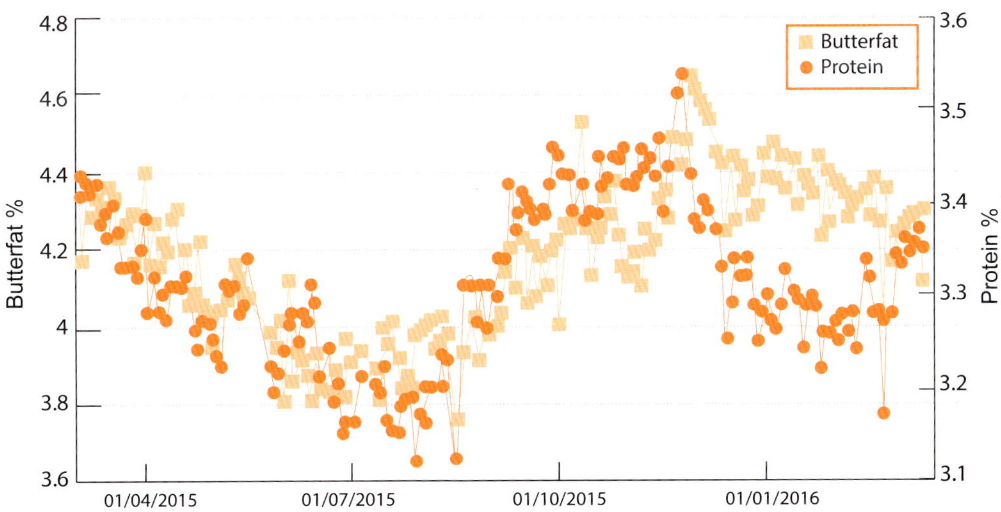

Fig. 26: Variations in butterfat and protein levels in a herd from NMR milk recordings

Pasteurisation

Pasteurisation is the process of heating milk for a target time at a target temperature. The goal is to reduce or eliminate disease exposure to calves.

There are two different methods for pasteurisation:

1. ***Batch pasteurisers*** *(Fig. 27)*

– 63°C x 30 minutes
– 38–700 litre batches
– 1.5–3 + hours depending on batch size
– An Agitator to allow even heating and cooling
– Automated heat and cool cycle
– Manual wash

2. ***Continuous flow pasteurisers***

– Flash or high temperature/short time
– 72°C x 15 seconds
– 5–180 litre per minute
– Automated heat and cool cycle
– Automated wash system

The rate of heat inactivation increases exponentially with time. However, pasteurisation should not be confused with sterilisation. Some heat-tolerant (usually non-harmful) bacteria will survive the process. Additionally, if poor-quality milk is pasteurised that already has a very high concentration of bacteria, then some harmful bacteria may survive the process.

The potential benefits from feeding pasteurised milk are:

1 **As compared to raw milk...**
– Improved health by reducing disease causing bacterial transmission

2. **As compared to a milk replacer**
– Improved rate of gain
– Improved calf health
– Improved longevity and future performance
– Improved economic efficiency

3. **Utilisation of a non-saleable product**

However, incorporating a pasteuriser onto a farm requires careful planning, with clear Standard Operating Procedures and thorough training of staff involved. Otherwise, there are a number of pitfalls.

Installation requirements[18]

a) *Hot water heater. Is a new one needed or is a heater self-contained in the unit? Does the existing hot water heater work? i.e. is the water hot enough?*

b) *Water supply*

c) *Are their special electrical requirements?*

d) *Space/location*

e) *Drainage requirements*

f) *Purchase and installation costs*

Fig. 27: A batch pasteuriser

Considerations for day-to-day use[18]

a) *Training farm staff to properly use and clean the equipment*

b) *Time/labour to use and clean equipment*

c) *Cleaning requirements*

d) *Variable costs*

e) *Service. Is the equipment reliable? How quickly can service be provided?*

f) *Moving and storing waste milk before and after pasteurisation*

g) *Monitoring performance. Is it working?*

Waste milk should be collected and stored in closed, clean containers, to prevent high bacterial contamination. Similarly, if milk is not to be pasteurised within a few hours of collection, it should be chilled in order to prevent bacterial growth and fermentation. Pasteurising fermented milk leads to curd formation, which deprives the calf of important nutrients for that particular feeding.[18]

After pasteurisation, any bacteria surviving the process will begin to replicate again if the cooling process is delayed. This can occur if the milk is allowed to cool slowly for several hours or is left to sit at warm temperatures for a very long period before being fed. All pasteurisers should therefore be equipped to rapidly cool the milk to feeding temperature immediately after pasteurisation is completed. Milk should then be fed as soon as possible after the process is complete. If there is to be a delay between pasteurisation and feeding, then the milk should be chilled.[18]

Post-pasteurisation contamination of milk is another concern. Pasteurised milk should be stored in clean, closed receptacles and distributed to calves in clean buckets or bottles. Careful attention must be paid to cleaning and sanitising buckets, bottles, nipples, etc.

With poor cleaning, fat, protein, and inorganic films (minerals) can build up in the pasteuriser itself: this interferes with temperature transfer to the milk and serving as a source to inoculate milk with bacteria. Clean this equipment as diligently as the milking parlour. Manufacturers or distributors of commercial on-farm pasteurisers should provide cleaning instructions that best fit the equipment.

Potential pasteuriser problems:[18]

1. *Start with poor quality milk with a high degree of bacterial contamination*

   – Chill stored raw waste milk to reduce incubation

2. *Milk not heated to the correct target temperature (Batch 63°C; HTST 72°C)*

   – Water heater doesn't get water hot enough or not enough hot water available

   – Inadequate plate cooler

   – Pasteuriser malfunctioning or not calibrated properly

   – Cleaning failure – build-up of fat, protein or inorganic films will interfere with heat transfer

3. *Milk is not maintained at the target temperature for a long enough duration*

   – HTST: milk not circulated for full 15 seconds

   – Batch: milk not kept at target temperature for a full 30 minutes

   – Operator error – if people are rushing they may stop the pasteurisation process before either target time or temperature is met

4. *Curdling of milk if fermented (acidic pH)*

   – Chill raw pre-pasteurised milk to prevent fermentation

5. *Post-pasteurised milk should be cooled rapidly (should be automatic in commercial machines) to prevent incubation*

6. *Post-pasteurisation contamination of the milk*

   – Store in closed, clean container. Chill if delay before feeding

Evaluating cleaning includes visual assessment for build-up of residual films plus performing monthly cultures of pasteurised milk (e.g. standard plate count, total bacteria count, lab pasteurised count).[18]

Monitoring pasteurisation equipment ideally requires the pasteuriser to be equipped with a time-temperature control chart to document that the target temperatures are being reached for an appropriate duration. At the very minimum they must be equipped with a thermometer by which producers can periodically check and monitor times and temperatures.[18]

Acidified milk feeding

Whole milk provides a favourable environment for bacterial growth. Low temperature and low pH slow down bacterial growth. Refrigeration of milk in a calf feeder is not practical, but lowering the pH can control bacterial growth and enable milk to be held without refrigeration for a short time. Preserving milk in this way allows larger quantities of milk for ad-lib feeding of calves. Levels of bacteria in milk will affect how long it can be stored before levels are reached that can affect calf health.

There is little evidence that acidification affects the nutrients in milk or milk replacer, or use of these nutrients by calves. Feeding acidified milk within 3 days minimises any protein degradation. Calf performance is not affected by the addition of acid. Formic acid kills 90% of *Mycobacterium* avium subspecies Paratuberculosis (Johne's disease) with 8 hours contact time at pH 4.0 and 100% with 48 hours contact time. There is, therefore, a potential for feeding whole milk in herds with Johne's disease issues.

The basic components of an acidified milk feeding system include a reservoir to contain the milk or milk replacer and a plastic tube with a one-way valve to carry milk to a rubber nipple (Fig. 28). Acidification with formic acid preserves the milk for storage at room temperature and allows mixing of batches at 1–3 day intervals.

Fig. 28: Plastic tube with one-way valve carrying milk from plastic container containing acidified milk to a rubber nipple

How to prepare formic acid 85% for use

- Formic acid 85% is hazardous to skin, eyes and lungs. Store acids safely and keep them out of reach of children
- Use goggles, gloves and an apron. Work in a well-ventilated area
- Mix 1 part formic acid 85% into 9 parts water (Table 7)
- Use a measuring cup with metric units for volumes
- Always add acid to water
- Label the container of dilute acid clearly (Fig. 29)

Fig. 29: Diluted acid labelled clearly. With measuring jug and clean equipment for transporting acidified milk

Preparing and feeding acidified whole milk

- Start with high-quality milk that has low bacterial contamination. Collect with clean equipment, cool immediately after milking and store in closed containers. Adding acid to milk will limit growth of bacteria but not kill what is already present
- Use cool (10–24°C) or cold (≤15°C) milk to avoid coagulation or clot formation
- Use dilute formic acid. Add 30 ml into each litre of milk (Table 8)
- Add the acid slowly into the milk and stir vigorously
- Stir again within the first hour, and three times each day for 10–15 seconds each time (Fig. 30)
- Check for target pH 4.0–4.5. Use narrow range (3.0–5.5) litmus paper or a pH meter
- It takes approximately 1–2 hours to kill coliforms; 4–6 hours to kill *Staphylococcus aureus* and 48 hours for Johne's disease.
- Feed at ambient temperature in summer and 20°C in winter
- Provide free-choice water and calf starter
- Clean nipples and valves daily, lines and containers weekly, with warm water and detergent
- Mount nipples at calf's shoulder height
- A ratio of 1:1 teats-to-calves is recommended
- Some calves may reject acidified milk, particularly in the first week of life. If calves are fed individually or young calves can be started in a separate group, diluting the acidified milk with whole milk may help to make the transition. Feeding from a nipple works better than an open pail

FORMIC ACID 85% (LITRES)	WATER (LITRES)	TOTAL DILUTE ACID (LITRES)
1	9	10
2	18	20
3	27	30
4	36	40
20	180	200

Table. 7: Diluting formic acid 85% 1 part acid into 9 parts water

DILUTE FORMIC ACID (ML)	WHOLE MILK (LITRES)
30	1
300	10
600	20
900	30
1800	60
3000	100
4000	133

Table. 8: Guidelines for mixing dilute formic acid into milk

Fig. 30: Barrels of stored acidified milk with day of feeding labelled clearly. The ingenious "homemade" mixer is an industrial paint-stirrer attached to a drill

It is possible to mix up enough acidified milk for 3 days. By 2 weeks of age calves will consume 8–12 litres per day. The acidified milk should not be exposed to heat or light. This has led to the development of many an ingenious way of housing feeding barrels – usually in plastic containers.

In summer the acidified milk should be fed at ambient temperature (provided this is not above 24°C). Above 24°C there are issues with milk separation and over 30°C leads to the formation of a hard cottage cheese-like curd. If the milk is fed too cold then the calf has to use too much energy to warm it up, so in winter keeping the milk uniformly warm is important. Use of a heating element as in Fig. 31 is one option, prior to dispensing acidified milk. However, if the milk is housed in a barrel for ad-lib feeding then use of alternative options are required. There is currently no method specifically developed for this, options used have included: fixed heating elements in milk, warming plates beneath barrels of acidified milk, warming of plastic pipes supplying milk to calves, water baths warmed by aquarium heaters, and many other homemade designs.

It is also possible to make up acidified milk replacer, with all the same benefits and constraints of whole milk.

Fig. 31: Heating element with a thermostat for warming acidified milk to desired temperature prior to feeding

Colostrum/transition milk

Excess fresh colostrum is an excellent feed for young calves. It is high in dry matter and protein content and low in lactose and may be stored refrigerated for up to 2 days (see Chapter 2, page 28). Storage and labour requirements may make its use as a regular calf feed prohibitive, unless a pasteuriser is available. It can also cause calves' faeces to become loose, which can be overcome by diluting 3 to 4 parts colostrum to 1 part water.[3]

Fermenting colostrum uses the surplus milk collected from cows during the first six milkings after calving – "Transition milk". This system is useful in large all-year round calving herds – where a continuous supply is available. Weight gains equivalent to feeding whole milk or milk replacer are possible. Table 9 shows suggested feeding rates. Occasionally, calves will not drink fermented colostrum. They should then be started on whole milk and gradually transferred to the fermented colostrum mixture.[3]

	DAILY AMOUNT (LITRES)	
BIRTH WEIGHT (kg)	COLOSTRUM	WATER
41–45	3.4	1.7
36–41	3.1	1.5
32–36	2.8	1.4
27–32	2.6	1.3

Table 9: Suggested feeding rate for fermented colostrum – diluted 2:1[19]

Liquid feeding systems

Milkfeeding systems most frequently employed in the UK are a bucket (36%) and a multi teat feeder (24%). Automatic feeders are used on 20% of UK farms.[1] The average volume of milk fed to a heifer calf per day is 4.58 +/– 1.33litres.[1]

There are almost as many milk feeding systems as there are farms in the UK! While this statement is a slight exaggeration, it is fair to say each farm varies slightly in how it feeds its heifer calves' milk.

The main systems are:

1. Restricted feeding systems

a. Once-a-day bucket feeding

b. Twice-a-day bucket feeding

2. Ad-lib feeding systems

a. Cold Ad-lib feeding

b. Automatic feeding

c. Computer controlled feeding

3. Accelerated or enhanced growth feeding systems

Fig. 32: An angled teat to encourage neck extension and oesophageal groove closure

Fig. 33: An open pail with the bottom raised 30 cm from pen bedding level

Restricted feeding systems

Restricted feeding systems are where calves are fed limited amounts (4–6L/day) of liquid milk, at the same time(s) each day. They were developed to encourage the calf to eat more starter feed at a younger age, limiting the costs of milk versus grain.

Calves are usually individually penned and fed via an open bucket/pail or via a teat attached to a bucket/pail. Milk from a teat encourages a more natural drinking behaviour, but it must be angled to encourage neck extension and effective oesophageal closure (Fig. 32).

- *The calf should keep her neck and head up and the height of the teat should equal the normal nose height of the calf*
- *The calf's nose should be below its eyes when drinking*
- *Its neck should not be fully extended*
- *Its head should be more horizontal than vertical*

An open bucket/pail should not be fed at ground level, but 30cm from pen bedding level[4] (Fig. 33). It is not a one-size-fits-all situation though:

- *Consider calf size and bedding depth when setting pail height*
- *Make sure no bars are pushing on the throat of the calf and that it can comfortably reach the bottom of the pail*

Once a day bucket feeding

Once-a-day feeding (feeding only one meal per day) is where milk is fed to calves once during a 24-hour period. It is a restricted feeding system with a very low labour input and low growth rates. The system requires a high level of stockmanship as calves are only seen drinking once every 24 hours. Additional viewing of calves, other than at feeding times, is required to check for signs of illness. If using milk powder the concentration is typically 15% (150g/litre). With calves fed 3–4 litres this equates to 450–600 g milk replacer/day.

Like all restricted feeding systems, once-a-day feeding aims to encourage starter uptake from an early age. Twice-a-day milk or milk replacer feeding is required by law (97/182/EC: Commission Decision of 24 February 1997 amending the Annex to Directive 91/629/EEC) until calves reach at least 4 weeks of age.[20] This is because a calf is required by law to have two feeds a day and a calf will not consume enough starter until it is 4 weeks old for this to count as one feed. So once-a-day bucket feeding cannot begin until calves are 4 weeks old. The pros and cons of once-a-day milk feeding are explained in Table 10.

PROS	CONS
Less labour due to less time spent mixing, feeding and washing up vs twice a day	Significantly lower opportunity for observation and supervision
Less milk replacer used	Time dependant, i.e. feed at same time each day
"Faster" uptake of concentrates	Can only be started at 4 weeks so requires a change in management around halfway through calf rearing period
Does not require power supply	Limited growth rates (300–400 g/day)
Can be offered via teat or open bucket	A lot of labour per calf
Individual weaning	Building space required will be more than group systems
Individual pen gives control on health and disease	
Relatively easy task to delegate	

Table. 10: Pros and cons of once-a-day milk feeding

Twice a day bucket feeding

Twice a day (see Table 11) is a restricted feeding system, more flexible and popular than once-a-day feeding. Calves are fed twice every 24 hours, at the same times each day. There is greater scope for increasing milk feed intake and growth rates. Additional viewing of calves, other than at feeding times, is required to check for signs of illness. If using milk powder the concentration is typically 10–12.5 % (100–125 g/litre). With calves fed 4–6 litres this equates to 400–750 g milk replacer/day.

PROS	CONS
Close control over volume fed, therefore, feed is cost effective	More labour intensive vs once a day
Improved supervision of calves vs once a day	Time dependant i.e. feed at same times each day
More flexible vs. once a day	Limited growth rates (400–500 g/day)
Does not require power supply	A lot of labour per calf
Can be offered via teat or open bucket	Building space required will be more than group systems
Individual weaning	
Individual pen gives control on health and disease	
Relatively easy task to delegate	

Table 11: Pros and cons of twice-a-day bucket feeding

Ad-lib feeding systems

Ad-lib feeding systems are where calves are provided milk on a continuous basis throughout the day and hence are more closely related to a "natural" situation. The systems were developed to reduce labour demand and encourage higher growth rates and feed conversion.

Calves are usually housed in groups, with design being important to manage the increased risks of disease. Calves spend about 45 minutes drinking milk each day, compared to a minute or two on restricted feeding. This can provide a natural outlet for sucking behaviour.[4]

Cold ad-lib feeding

This system is uncomplicated and basic as milk is provided in a large container and fed via tubes to a teat (Fig. 34). This allows several calves to access feed at the same time. High levels of stockmanship are required as the usual cues of whether a calf has fed or not will not be present compared to controlled feeding times.

Fig. 34: Calves feeding from a cold ad-lib system

Feed milk cool to avoid gorge feeding, and provide one teat per calf. There is no automatic washing of teats and so they represent a source of disease spread among a group. Disinfect teats twice daily and then wash with warm water. Alternatively rotate teats on a daily basis – one set in disinfectant, while the other set is in use - then swap after 24 hours. This is good hygiene, and will prolong the life-span of teats. Damaged teats increase the risk of aspiration pneumonia – monitor their condition and replace regularly.

Milk in the container is subject to environmental temperatures – too cold and growth rates are reduced, too hot and bacterial growth in milk could lead to disease. Acidification can be used to preserve milk for longer (see page 56).

If using milk powder the concentration is typically 10–12.5 % (100–125 g/litre). With calves able to consume around 12 litres/day ad-lib, this equates to 1200–1500 g milk replacer/day. This system therefore usually uses whole milk (surplus/waste/transition/ mixture), due to the potential high cost of milk replacer.

PROS	CONS
Low cost equipment and set up	No ability to tell which calves drink and how much they take – stockmanship levels must be high
Low labour cost	Lower hygiene standards = increased disease risk
Freedom from specific feeding times	Greater urine production increases bedding requirement
Does not require power supply	Higher milk replacer consumption vs restricted feeding systems
Relatively easy task to delegate	Starter intake limited at all stages
Fast growth rates possible	Management difficulties in cold and hot weather
Allows 24 hour milk consumption	
Can use surplus/whole milk	

Table 12: Pros and Cons of Cold ad-lib feeding

Automatic Feeding

This system can considerably reduce the time and labour needed to feed calves (See Table 12). It allows calves to access feed that is consistent in temperature and concentration, 24 hours a day. High levels of stockmanship are still required as the usual cues of whether a calf has fed or not will not be present compared to controlled feeding times.

Milk replacer is mixed with warm water at a set rate. Calves are fed via a teat with a self-closing hole (Figs 35 & 36). Teats should be 1 m apart and 60–80 cm from the ground.[4] One teat per 7–8 calves. Calves will require training to the feeder – see below (Computer-controlled feeders on p63).

Some machines have a self-cleaning facility that provides a way to maintain sanitation of the unit. It should not be wholly relied on though, as machinery should be checked regularly to ensure it is in good working order and there is no build up of residues anywhere. Cleaning of the mix bowl and pipes weekly is essential. Teats will still require disinfecting twice daily and then washing with warm water. Alternatively rotate teats on a daily basis – one set in disinfectant, while the other set is in use–then swap after 24 hours. This is good hygiene and will prolong the lifespan of teats. Damaged teats increase the risk of aspiration pneumonia – monitor their condition and replace regularly.

Fig. 35: An automatic calf feeder

Fig. 36: Teats from an automatic calf feeder

PROS	CONS
Allows 24-hour milk consumption maximising early calf growth potential	No ability to tell which calves drink and how much they take – stockmanship levels must be high
Consistent feed supply – temperature and concentration	Greater urine production increases bedding requirement. Drainage must be good in the feeding area (1 in 10 around drinking area)
Low routine labour requirements	Higher milk replacer consumption vs restricted feeding systems
Flexible and portable	Starter intake limited at all stages
High growth rates	Hygiene standards can drop due to a reliance on the machine
Suitable for groups up to 20 calves	Training calves at entry requires time and patience
	A reliable, up-to-date power supply with back up is needed and a clean supply of water under constant pressure
	Initial high cost but this can be offset against labour and performance savings

Table 13: Pros and cons of automatic feeding

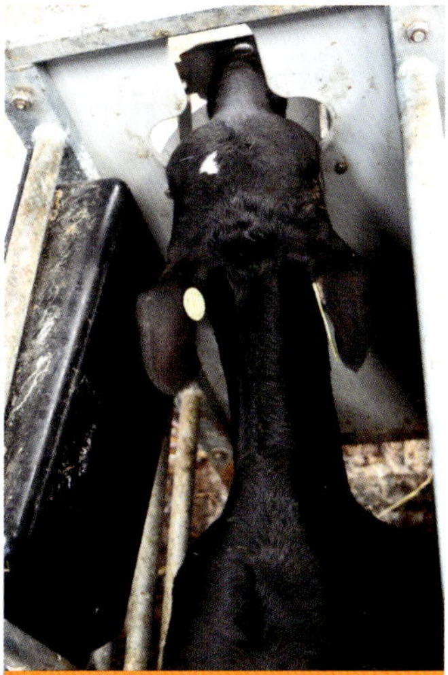

Fig. 37: Calves wear a transponder – either an ear tag or a collar. In this calf it is located within the left button ear tag

Computer controlled feeding

This system aims for ad-lib performance with the cost control of a restricted system. Calves wear a transponder (ear tags or collar) that responds to a sensor at the milk feeding point (Figs 37 & 38). A computer identifies the calf within a group; and a machine dispenses a controlled amount of feed, a controlled number of times in a day (Figs 39 & 40). The milk is freshly prepared for calves at a set concentration and temperature (mixed at 40°C, fed at 37°C)[4].

Computer-controlled feeders can supply up to 4 feeding stations/teats and so are able to feed multiple pens of calves. Manufacturer recommendations are up to 30 calves per feeding station/teat - this is too many. Do not have more than 20 calves in a pen or the risks for disease are too high, and it is also unlikely each calf will be able to receive its daily allocation of milk.

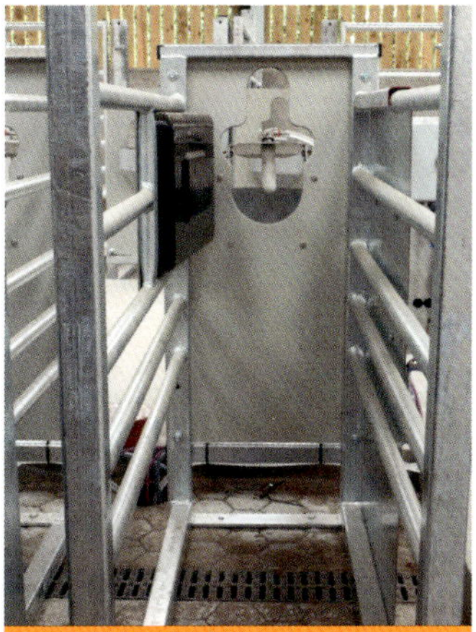

Fig. 38: The transponder in a calf's ear is read by a sensor in the black box mounted on the left-hand side of the feeding station

Fig. 39: Milk powder is stored in the hopper and this is dispensed and mixed at a set concentration and temperature before feeding to the calf

Fig. 40: Individual calf identification allows each calf to receive a tailored amount of feed for each meal and each day

Calves will require training to the feeder:

- *Skip the morning feeding, place the calf in the automatic feeder/computer controlled feeder pen and lead to the nipple*
- *90% of calves will learn in 2 trips*
- *10% need more assisted trips*
- *Don't overdo it (kill them with kindness), or calves will associate a "person presence" with being fed*
- *It takes an average of 1.5 days to train a new pen*
- *Calves visit the feeder less than 8 times a day with an average of 5 meal visits and 3 visits where they don't drink milk*
- *Weak calves should not be introduced to the feeder and group until they are strong enough to drink on their own*

The computerised feeders have the added benefits that the computer records intakes – meal times, duration, volumes. This can aid in the detection of sick calves but runs the risk of an over-reliance on the computer to identify sick calves. Computers are poor substitutes vs good husbandry. Individual medication and electrolyte administration programmes are also available. The computerised feeders also reduce milk replacer available to the calf gradually over the last few weeks before weaning (Fig. 41).

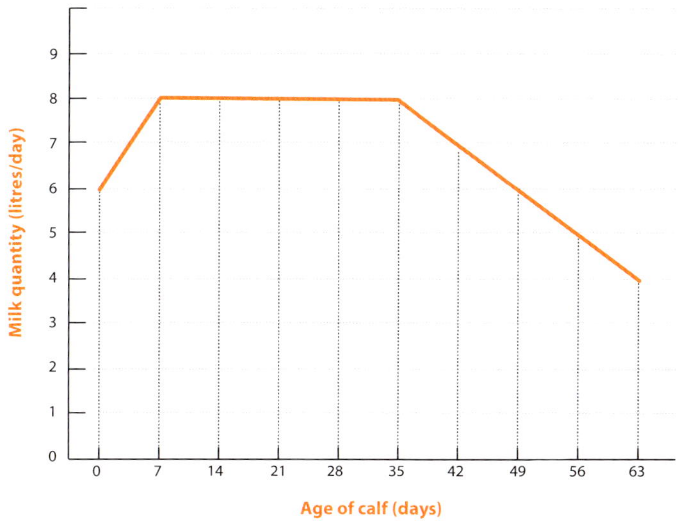

Fig. 41: An example of a feeding programme for calves on a computer-controlled feeder

Machines have a self-cleaning facility that provides a way to maintain sanitation of the unit. It should not be relied on though, and machinery must be checked regularly to ensure it is in good working order with no build up of residues. Clean the mix bowl and pipes weekly. Teats require disinfecting twice daily and then washing with warm water. Alternatively rotate teats on a daily basis – one set in disinfectant, while the other set is in use – then swap after 24 hours. Good hygiene will prolong the lifespan of teats. Damaged teats increase risk of aspiration pneumonia – monitor their condition and replace regularly.

PROS	CONS
Control of feeding programme to calf/farm needs	Good housing essential
Each calf fed to individual needs	Greater urine production increases bedding requirement. Drainage must be good in the feeding area (1 in 10 around drinking area)
Computer analysis of individual or groups of calves daily feed consumption	Higher milk replacer consumption vs restricted feeding systems
Consistent feed supply – temperature and concentration	Stockmanship and hygiene standards can drop due to a reliance on the machine
Low routine labour requirements	Training calves at entry requires time and patience
High growth rates	A reliable, up-to-date power supply with back up is needed and a clean supply of water under constant pressure
Individual graduated weaning programme	Initial high cost but this can be offset against labour and performance savings
Self-cleaning (but requires monitoring)	

Table 14: Pros and cons of computer controlled feeding

Accelerated or enhanced growth feeding systems

This system aims to feed calves so they develop according to genetic potential. Accelerated growth milk replacers have 26–30% protein and 15–20% fat. Whole milk has 27% protein and 30% fat. So protein in accelerated growth milk replacers is similar, but fat is less – with the aim of promoting lean tissue gain and efficiency of gain. An improved immune system with reduced scour and pneumonia is an additional benefit.

Feeding from birth to weaning

Calves can be housed individually or in groups, and fed via a bucket or machine. Feed calves 2.5 litres twice daily in their first week of life, then 4 litres twice daily from week two up until weaning. When using an accelerated growth milk powder, concentration is typically 15% (150 g/litre). With calves fed 8L per day from week two this equates to 1.2 kg milk replacer/day.

A higher protein calf starter (22–24% DM) is also required to complement and maintain growth levels. Intake of starter lags behind calves fed on conventional systems, but increases at approximately the same rate once the amount of liquid is cut back (see Weaning page 72). Calves must have access to free choice water from birth in this system. Calf faeces will be softer than normal and larger in volume.

To make these feeding programmes cost effective, the increased cost of high protein milk replacer and the extra cost to feed more dry matter must be offset by long-term improvements in growth or decreased overall heifer production costs – they invariably are.

PROS	CONS
Increased growth rates from birth until weaning	Greater urine production increases bedding requirement. Drainage must be good in the feeding area (1 in 10 around drinking area)
Increased lean tissue to fat tissue ratio	Higher milk replacer consumption vs restricted feeding systems
Increased efficiency of gain	Relatively high cost of milk replacer, but this can be offset against performance savings
Increased mammary tissue in the udder for more potential future milk production	Weaning can be later at around 8–10 weeks vs. restricted system of 8 weeks
Increased immune response (decreased sickness and death) death loss <1%	Weaning management is important to ensure sufficient rumen development
Decreased medicine costs up to 80%	
Decreased age at first calving	
Programme does not have to be altered depending on environmental conditions	
Increased 1st lactation milk production around 800 litres	
Flexible system used for individually housed calves or groups; bucket fed or machine fed	

Table 15: Pros and Cons of Accelerated or Enhanced Growth Feeding Systems

Which liquid feeding system do I choose?

Nutrition and management practices applied to today's dairy replacement heifers will influence the performance and economic returns of the dairy herd in 2 years time. The answer to how you decide to feed your calves today depends on how you want your dairy herd to perform in 2 years time and beyond?

The calf left on its mother or fed ad-lib will consume 20% of their body weight or 10–12 litres a day. They can easily achieve 1kg of daily weight gain. Restricted feeding systems grossly undersupply the calf's needs and restrict potential. Starter intake is negligible in the first 3 weeks of life. Therefore, calves on restricted feeding regimes are at most only able to achieve 20–30% of their biologically normal growth.[21]

The main barrier to feeding calves greater than 2 litres twice daily (in a conventional restricted system) is likely to be cost, however, an increase in the size, or frequency of milk feeds, is beneficial to growth and immunity and cost effective in the long term if heifers attain target weights and enter the milking herd earlier.[22]

For every 100 g of average daily gain in the first 2 months of life (mainly achieved by increases in milk or milk replacer consumption), an additional 225 kg of milk can be expected during the first lactation.[23] Increased growth rate before weaning has positive effects on lactation milk yield. The effect of growth rate and thus nutrient intake prior to weaning has a more direct and significant effect on milk yield than genetic selection for production. Genetic selection yields between 68–115 kg milk per lactation. Pre-weaning calf nutrition and management can yield 4 to 8 times more milk than genetic selection per lactation. Furthermore, increased growth during the first 2 months increases a heifer's chance of surviving to second lactation.[24] Bigger, better calves at weaning perform better post-weaning.

The dairy animal is at its most feed conversion efficient when it is a pre-ruminant calf. Maximising these efficiencies for the longer-term benefits of higher lactation yields, reduced disease and prolonged lifespan in the milking herd must be taken advantage of. This means feeding more high-quality milk replacer or whole milk, and not using restricted feeding systems. The modern dairy farmer should not be aiming to make cost savings with calves, as the investments at this stage will be easily repaid, and lead to greater long-term benefits.

Cold weather feeding strategies

Calf maintenance requirements, but not crude protein requirements, increase during cold weather. To maintain health and growth under these conditions requires increased nutrient intake. Several strategies can be used:[6]

1. *Increase the volume of milk or milk replacer at each feed (Table 2, page 41)*
2. *Provide a third feed daily*
3. *Switch to higher energy (higher fat) milk replacer, e.g. if feeding a low fat (15%) milk replacer switch to 20% fat in the winter months*
4. *Supplement milk replacer with added fat*
5. *Add additional milk replacer solids to each feed to increase the concentration (no greater than 160 g/litre). Free choice water must be available at all times*

Calf starter

Calf starter is the first solid dry feed fed to calves. It is a complete feed, made from grain, protein sources and fortified with vitamins and minerals, and is available in either a pellet or coarse feed (Figs 42 & 43). Calf starter pellets are most popular, and fed on 60% of UK dairy farms.[1] Calf starter must provide:[4]

- *Stimulation for early rumen development*
- *Suitable presentation and palatability for early intakes*
- *Balanced nutrition to supplement calves liquid diet for growth, skeletal development and a strong immune system*

Fig. 42: Calf starter pellets

Fig. 43: Coarse calf starter

As calves begin to consume dry feed, the rumen develops a microbial population and absorptive function. The source of nutrients begins to change from products absorbed directly from digestion of milk or milk replacer to a combination of dietary ingredients and end products of microbial fermentation (Volatile Fatty Acids (VFA), microbial protein).

Fermentation of dry feeds to the VFA butyrate and propionate is necessary to drive growth and differentiation of the rumen lining. Good calf starters allow high rates of fermentation and microbial protein manufacture, yet still provide some bypass protein and starch to be digested in the lower tract[6]. With a good quality starter plus milk then a calf should start rumination at 5–6 weeks of age. Any earlier than this and the calf is not getting enough milk.

Starter is best presented as a coarse feed or pellets. The grains included in the calf starter should be coarsely ground, pre-heated, cracked or rolled to increase palatability and stimulate rumen development. Feed processing can have effects on rumen development, e.g. steam flaked cereals can increase rumen propionate.[4]

Calf starter freshness is important – it must smell and taste good. Starter should not be dry, dusty, mouldy, or have an "off" flavour. Very fine-textured feed tends to cake together when wet, resulting in low intake. High-moisture grains are not recommended, since they heat quickly and mould in buckets.[3] The growth rate of young calves depends heavily on grain intake. Unpalatable or poor-quality starter grains inhibit intake, reducing rumen development and decreasing growth.

Find out the formula of any starter mix before you feed it to determine if it's any good! Minimum crude protein (CP) levels in calf starter are 18%. Increasing to 22–24%, promotes lean tissue and growth and reduces quantities needed to be fed when it is time to wean. The best protein source is soybean meal (an absolute no–no in milk replacer).

The main proportion of energy in starters is from cereal grains – starch (maize, oats, barley, wheat) for microbial growth and production of VFAs, and sugar (molasses – minimum 7%), to encourage intake, reduce separation and waste. Fibre for rumen development and stimulating intakes is usually from sugar beet pulp, wheatfeed, grassmeal, dried brewers grains, etc.

Starters normally supply the minerals and vitamins required by the calf in adequate amounts. As additives, several compounds are approved for use in starters to control coccidiosis. As the effective dose depends on projected intake of starter dry matter, lower intakes may make the compound ineffective.

Fig. 44: Overfull bucket of calf starter – old, with wet compacted material around the edges, and highly unpalatable

How to feed:

- *Offer fresh starter every day, from 4 days to 4 months of age*
- *Start by placing a small handful in a bucket each day after morning milk feeding if on buckets, or in a trough each day if in groups*
- *Use a shallow bowl to bring the height of the grain up for very young calves and allow them to lick the bottom. Once the calf is comfortably eating starter this can be replaced with a normal sized bucket (around 2 weeks)*
- *Increase amounts gradually in response to calf intake up to:*
  - *1 kg starter/calf/day if CP ≥ 22%*
  - *2 kg starter/calf/day if CP <22%*
  - *This requires accurate weighing of calf starter*
- *Remove any starter that is left at the end of the day and feed to older calves*
- *Keep feed containers clean (daily) and free of wet compacted material – if you left your breakfast cereal yesterday – you wouldn't really want to eat it today! (Fig. 44)*
- *Feeding equipment must be in good condition and appropriate to the calves being fed*
- *Store feed in a dry covered bin to limit moisture and access to birds and rodents*
- *Ensure plenty of fresh clean water is available as soon as starter is offered*

Forage

Forage is fed to heifer calves on 76% of UK farms, most frequently as wheat straw (46%) (Fig. 45), followed by hay and barley straw (11%).[1] Whether or not to feed forage to calves pre-weaning is controversial. This arises from the idea it is natural for calves to consume forage. Cattle developing under wild conditions, however, undoubtedly calved in the spring when grass was immature and abundant. The first solid feeds available to the young calf then would be fresh immature grasses – rich in sugars and low in fibre. The high sugar content fermented to a relatively high proportion of butyrate and propionate – ideal to stimulate rumen papillae development. Contrast this with forages offered to housed calves that are low in sugars and higher in poorly fermented fibre.[6]

Fig. 45: Calves eating wheat straw

Limited body size means calves have a high energy requirement relative to their ability to consume dry feed. Therefore, if calves consume significant amounts of forage, intake of starter will be limited. If starter intake is reduced, growth is reduced. Nutrients in starter are more rapidly and completely digested than those in forages. The complex carbohydrates in forages reduce digestibility rates in young calves. Most forage has too little energy per unit of feed for calves. When rumen microorganisms digest forages, more acetate is produced. This will not allow maximum rumen papillae development compared to starter, which is fermented to butyrate and propionate. In addition, calves have limited space in their digestive tracts. Bulky forages quickly fill this space, sending a signal that depresses appetite. Thus, gut-fill quickly becomes a limiting factor to obtaining enough nutrients.[3]

Don't feed pre-weaning calves forage. If a calf is receiving higher quantities of good-quality milk powder, for example, in an accelerated milk feeding programme, coupled with high protein starter (22–24%) protein – they won't want to consume forage. Their appetite will be satisfied. If not on this type of feeding programme it is very likely a calf will consume forage if offered, or in its bedding.

If calves are housed in groups and there is a scour outbreak – consuming bedding is an excellent way to spread the disease. If a calf is housed individually and has scour – consuming bedding is an excellent way to re-infect itself. In these circumstances having a limited amount (150–200 g per day) of high-quality, chopped, digestible forage is a preferred alternative to encouraging disease spread. Otherwise stick to milk replacer and starter until a calf is weaned.

Water

Provide calves access to water from birth. It is an essential nutrient for all animal life and will make up approximately 75% of a calf's body weight. There are two key areas to encourage water consumption:

1. *Quality – water should be high quality and kept fresh*
2. *Quantity – enough water should be offered to promote healthy growth and rumen development (Table 3, page 42) (approximately 4.5 litres/kg DM intake)*

Water is important for many basic metabolic functions in the calf's body. For feeding calves, water is important for:

- *Rumen development – calves are born without a functioning rumen, which is developed by fermentation of starter and water*
- *Improved starter fermentation – milk replacer consumed by the calf is directed to the abomasum by the oesophageal groove. Water is directed to the rumen and aids fermentation of calf starter, plus it is essential for bacteria and protozoa - responsible for this fermentation, to survive in a water environment in the rumen*
- *Better starter intake – calves offered free-choice water consume more calf starter and so grow faster, and have a stronger immune system*
- *Calves that scour will increase their water intake voluntarily thus improving their chance of recovery*

Water must be provided:

- *Clean – it can become contaminated with chemicals, microorganisms and high levels of minerals. Dirty water reduces intake and subsequent starter intake, growth and health (Fig. 46). A sign quality may be suboptimal is when calves start lapping the surface water, but failing to drink. Provide fresh clean water twice daily*
- *A cool (10ºC) is best for performance and it allows the calf to get rid of excess heat (calves cannot sweat but need to stay cool).* [4] *In cold conditions though, warm water should be provided to reduce "cold load" on the calf, and increase time to water freezing*[4]
- *Via a clean bucket, trough or automatic system. Action a regular maintenance and cleaning routine (twice daily) to maintain hygiene and adequate flow/supply*
- *With sufficient access so as not to limit intakes*
- *Kept physically and "mentally" separate from milk feeding, e.g. for a "two-bucket" pen, milk and starter pellets should be swapped at feeding, not milk and water.*[14] *Separating water and dry feed physically or with dividers stops calves slopping water into the dry feed (Figs. 47–49)*

Water-quality issues, such as the presence of elevated minerals or contamination with microorganisms are a good reason to regularly test the water given to calves. Particular attention should be paid to well water sources and especially after snow has melted, heavy rainfall and excessive run-off situations.[4]

Fig. 46: Dirty water reduces intake and subsequent starter intake, growth and health. If you would not drink from this bucket, then why should a calf have to?

Fig. 47: Placing starter and water buckets next to each other can lead to slopping of water into the starter – Fig. 48

Fig 48: Water that has slopped onto the starter has broken it down and made it unpalatable

Test water quality every 6 months, submit samples in sterile containers for analysis, and ask for an assessment of:

- Total dissolved solids – providing an indication of both the total and suspended inorganic matter in the sample. High concentrations of dissolved solids – such as sulphate, chloride, iron, manganese and nitrates can affect performance. In particular, high iron levels can result in reduced palatability, immune dysfunction and decrease absorption of copper, manganese and zinc from the diet[25]
- Hardness – calcium, magnesium, sodium, chloride and sulphates
- Heavy metals
- Bacterial contaminants – *E. coli* and Salmonella in particular

Fig. 49: Water and starter clearly separated to prevent water slopping onto the starter

Weaning

Weaning in calves is stopping feeding milk. It is the stage of heifer rearing when the rumen has achieved adequate development so that the calf can support its nutritional needs from rumen fermentation and digestion of dry feed.[4] Weaning allows a reduction in feed and labour costs compared to milk-fed calves. In the UK weaning occurs on average at 62 days, with the most common criteria for deciding to wean being age (53%), size (19%) and feed intake (17%).[1]

Fig. 50 Weigh calf starter accurately

Weaning represents the second most stressful event in a young calf's life. Failure to manage weaning well can result in stalled growth rates and potentially contribute to outbreaks of disease. A total of 90% of dairy herds experience a weaning dip, with calves failing to grow between 2 and 3 months of age.

Weaning depends very much on the milk-feeding programme and quality of calf starter. Computer-controlled feeding has a built in programme to encourage starter intake and allow weaning at a set time (Fig. 41). This still requires careful monitoring, and like all milk-feeding programmes, weaning should be based on calf-starter intake. Do not wean on the basis of age or size.

How?

- *Base weaning on calf -starter intake*
  • *If on a 23–25% protein starter the calf should be eating 1 kg of starter per day consecutively for 3 days*
  • *If on a conventional starter: 18–22% protein, the calf should be eating 2 kg of starter per day consecutively for 3 days*
  • *If the calf eats the desired amount for 2 days and then reduces the amount eaten, start over*
  Accurate recording of starter weight is required for this (Fig. 50) plus accurate recording of when calves reach the desired intake. This can be achieved in a diary, on a notice board or more simply using a colour-coded system (Fig. 51)
  • *In group-reared calves individual intakes are not measurable. Use the average amount eaten by the group before commencing weaning. So a group of 10 calves, on a 20% protein starter, must be eating 20 kg daily for 3 consecutive days, before commencing weaning*
- *Once starter is being consumed in this way reduce volume of milk fed by half. This is most easily done by removing one feed, or halving volume of milk fed per feed*
- *Feed half the volume of milk for one week, and then stop altogether. During this time increase starter fed in response to calf appetite*
- *Ensure fresh, clean water is available at all times*

This weaning "transition period" should be treated in the same way as an adult cow's transition period: from dry cow – transition cow – milking cow. The weaning transition takes place over 4 weeks (2 weeks before and 2 weeks after). If weaning calves based on intake, they will be much more consistent in size, and transition will run smoothly with minimal disease and growth problems.

Fig. 51: The use of coloured clothes pegs on the roof of this hutch clearly identifies how many days the calf has been consuming the desired amount of starter for, before commencing weaning. This system can also be used for days of treatment for an illness

Pitfalls

- *Do not base weaning on age or size*
- *Reducing concentration (dilution) of milk fed is not recommended*
- *Avoid weaning at times of stress – hot and cold weather, moving or mixing calves, de-horning, gathering for vaccination, etc.*
- *Avoid weaning if calves are showing signs of disease such as scours or pneumonia*
- *The post-weaning dip – often seen as reduced growth for 1–4 weeks after weaning, is entirely avoidable if weaning is managed properly*

Afterwards

- *Continue to provide the same starter as pre-weaning on an ad-lib basis for 7 days. Don't move onto a rearer until the calf is at least 4 months old*
- *Provide high-quality forage to encourage growth of rumen muscular layer and health of the rumen lining*[4]
- *Leave calves in their individual pens or groups for 7 days after weaning to avoid additional stresses of changes of environment and group competition. Then place individually housed calves into groups of four to six calves housed together in a separate pen*
- *Put the transition groups together after a further 2 weeks – provided there are no health problems. Group-reared calves should be placed into larger groups 2 weeks after weaning*
- *Separate any changes in feed or routine by one week*

Feeding programmes

Calves like a boring life. They want the same feed, in the same way, at the same time, every day. Any variation increases stress for the calf, reduces growth rates and increases risk of disease. A feeding programme is a plan of activities and actions that allow this consistency in calf feeding. It requires routine. It must be clear and straightforward, regularly monitored and adaptable. It should be designed to meet the heifer-rearing targets of the farm.

The feeding routine must set out the timing of feeding and the daily amount of milk to be fed from birth to weaning. It sets out the types and amounts of each feed, and the number and timings of feeds per day. [4]

Have clear guidelines that everyone involved with calf feeding understands.

Anyone who is involved in the calf feeding must be aware of the routine. There should be clear lines of communication between staff such as a diary, a notice board in the calf shed and calling, texting or e-mailing to pass on relevant information.

The routine should also include regular assessment of the success of the feeding programme, which will include monitoring performance and record-keeping. Performance will focus on growth and disease incidence.

Feeding Guidelines

1. Milk quality – whole milk or a high-quality milk replacer

2. Milk temperature – feed at a consistent temperature of 37-39 °C for both whole milk and milk replacer. If mixing milk replacer do not use hot water (>45°C) as this will damage the proteins in the milk. Use a thermometer to check temperature

3. Milk volume – feed consistent volumes daily. For bucket-fed calves, volumes should not vary from feed to feed. Any changes to amounts fed should be done on weekly basis, and there should be no other management changes done at the same time

4. Milk concentration – mix milk replacer at the same concentration every time, e.g. 15% solids – use 150 g of milk powder plus 850 ml of water to make 1 litre of mixed milk

5. Mixing milk – use a stainless steel whisk. Milk replacer should be mixed thoroughly until all solids are dissolved

• Under-mixing leads to clumping

• Over-mixing leads to disruption of fat suspension

6. Time of feeding – feed calves at the same time each day. If feeding twice a day, it doesn't have to be exactly 12-hour intervals; you may feed at 7am and 5pm, but this should be consistent. If feeding once a day, stick to 24-hour intervals rigorously

7. Feeding height

• When drinking from a teat the calf should keep her neck and head up and the height of the teat should equal the normal nose height of the calf. The calf's nose should be below its eyes, its neck should not be fully extended, and its head should be more horizontal than vertical

• An open pail must have its bottom raised 30 cm from pen bedding level

8. Feed equipment and areas – store, prepare and feed milk in hygienic conditions. Clean feeding equipment after every feed; do not use the same utensils for sick and healthy calves. Maintain feeding equipment (physical and mechanical) weekly. Calibrate weighing equipment/automatic feeding machines weekly

9. Calf starter – use high quality, from 4 days to 4 months of age. Offer fresh daily and dispose of any uneaten starter, which can be fed to older calves. Increase amounts given gradually in response to appetite

10. Fresh water – must be clean, plentiful and available at all times from birth. Monitor quality every 6 months

11. Have consistent personnel feeding calves

12. Observe - observe calves during the day for feed refusals, stool consistency, signs of illness, etc

Monitoring growth

Feeding the heifer well to weaning means she is strong and healthy post-weaning. The target beyond this is for her to conceive and then calve into the milking herd as a strong, healthy, productive animal. The ideal age of first calving is 24 months, in order to hit this target the heifer must be bred by 15 months, and so be of suitable size by 13–14 months. It is weight rather than age that determines when the heifer first comes into oestrus.

Benefits

- *Monitoring from an early age ensures calves meet targets for conception*
- *It allows good growth in early life when feed conversion is most efficient*
- *Identify underperforming and sick calves*
- *Identify problems in the management system*
- *Maximise growth efficiency and so be cost effective*

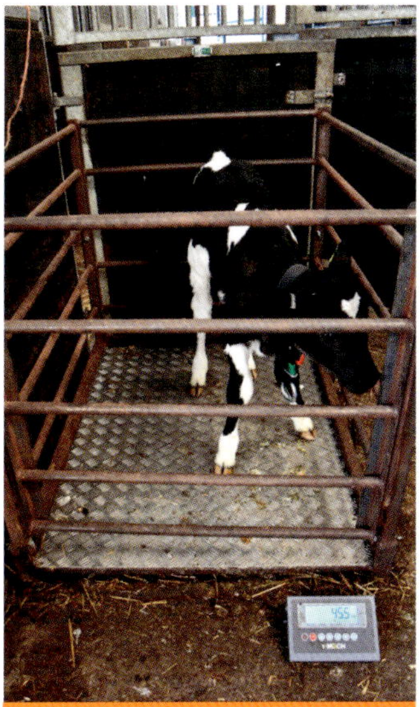

Fig. 52: Electronic weigh scales

How?

A weigh tape or electronic scales. Scales are more accurate but either method is suitable provided it is consistent – make sure the same person is carrying out the weighing each time.

Electronic scales (Fig. 52)

- *Dedicated calf scales or set up with load bars and a strong platform*
- *Encourage calves to enter using a race*
- *A good handling system is needed for scales to work successfully*
- *Calibrate monthly and/or check for accuracy*

Fig. 53: Weighing a calf with a weigh tape

Weigh tape (Figs 53 & 54)

- *Most accurate up to weaning*
- *Ensure calf is standing normally with head forwards and legs comfortably together*
- *Wrap the tape around the chest behind the front leg and the shoulder blade*
- *Keep the tape flat against the calf's body*
- *Have consistent tightness*
- *It is much easier with two people to handle the calf*

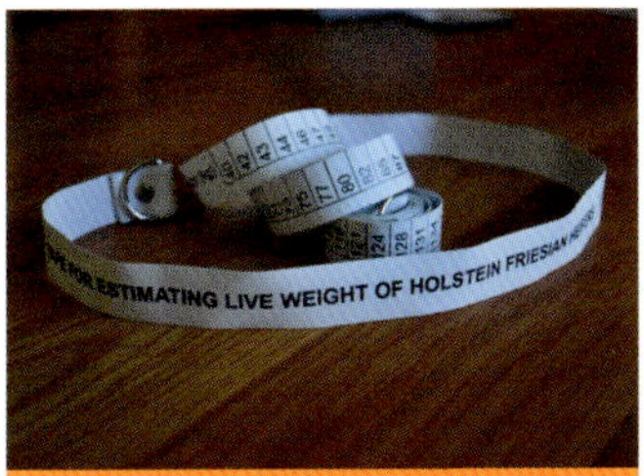

Fig. 54: A weigh tape

Feeding from birth to weaning

When?

The more information available the better informed you are. Regular, consistent measuring is very important. It can be performed at the same time as other tasks such as disbudding, vaccinating, moving, etc. but set times are important. Two measurements at different times allow growth rate to be determined for an individual calf. The best times are:

- *Birth – to give a baseline*
- *Regular intervals thereafter, e.g.*
  – Weaning – to look for a dip
  – 6 months of age – to allow time to make corrective changes before breeding if necessary

Average daily weight gain is then calculated by:

$$\text{Average daily weight gain} = \text{Weight gain} \; \frac{(\text{Latest weight} - \text{previous weight})}{\text{Number of days between recordings}}$$

Example:

$$\text{Average daily weight gain} = \text{Weight gain} \; \frac{(75-45)}{50} = 0.6 \text{ kg}$$

An alternative method is to take a "Snapshot":

- *Weigh all the calves of different ages on the same day, e.g. from 0 to 100 days*
- *Calculate the average weight of a newborn calf from those you weigh on the day*
- *As in the above calculation the average daily weight gain for each calf can then be calculated assuming each has the same estimated birth weight*
- *Age can also be plotted against weight* (Fig. 55)
- *This will allow individual calves to be highlighted as doing well or poorly, plus how the feeding programme is working as a whole*
- *Taking a "snapshot" is a good starting point but regular consistent weighing is better long-term*

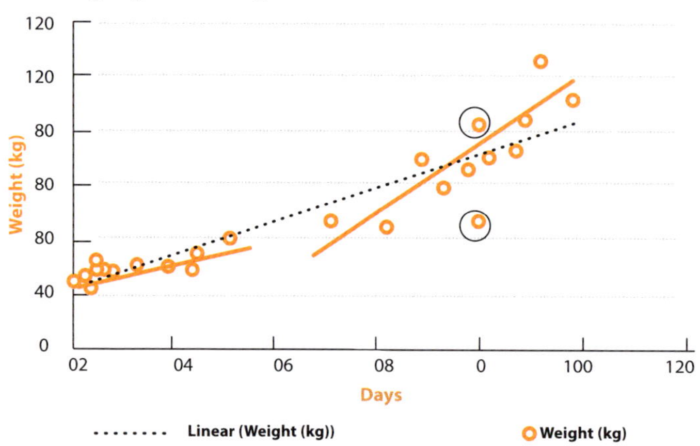

Fig. 55: A "Snapshot" of weight for 29 calves all weighed on the same day. The graph shows that weight gains are very slow in the first 3 weeks of life but then accelerate (orange lines). The graph also highlights the differences between individuals (circled) of the same age but significantly different weights – disease or management?

Table 16 shows suggested target body weights for Holstein-Friesians. Calculating the optimum growth rate is breed dependant and possible for an individual herd:

- *Weigh six 3rd lactation cows and average their weights (these cows are considered fully mature)*
- *The heifer wants to be 55–60% of adult body weight (about 13–14 months) at breeding. If the adult cow is 660 kg, then at breeding the heifer wants to be 360–400 kg*
- *If calf birth weight is 40 kg then the average daily weight gain per day up to 13–14 months of age is 0.75–0.8 kg*

AGE (MONTHS)	TARGET WEIGHT (KG)	% OF MATURE BODY WEIGHT
Birth	36–40	6
1	55–60	8
3	95–110	15
6	180–200	28
9 (Puberty)	274	40
15	375	55–60
22–24 (First calving)	582	85
Mature body weight	685	100

Table 16: Suggested target body weights for Holstein-Friesians

Growth targets will be "defined" by the breed of cattle, feeding and management systems available. Growth targets will be "achieved" dependant on the type and amount of liquid feed, management capabilities and health status of the rearing unit.[4]

- *Feeding management – ad-lib systems and accelerated feeding programmes have higher growth rates than conventional systems (Table 17)*
- *Health management – sick calves and calves with poor immunity do not grow well*
- *Breed – birth weight influences growth rate from feeding rate. Energy requirements increase for larger calves*
- *Genetics – high genetic merit animals have higher growth and production potential*
- *Sex – uncastrated bulls have a 12% advantage over heifers*[4]
- *Temperature – maintenance requirements increase in temperature extremes with a negative effect on growth rates*

Feeding and overall management systems cannot be assessed using growth rates alone. Disease and death are two obvious further measures - see Chapter 7.

PROGRAMME	45 KG CALF, MILK DRY MATTER INTAKE (G)	EXPECTED AVERAGE DAILY GAIN	
		0–21 days	21–42 days
Conventional milk feeding (2 litres twice daily)	400–500	0.2–0.3	0.5–0.6
Accelerated milk replacer	900–1000	0.5–0.6	0.6–0.8
Moderate	700–800	0.4–0.5	0.55-0.65

Table 17: Expected growth rates under different milk feeding programmes

References

1. Boulton, A. C., Rushton, J. and Wathes, D. C. (2015) A study of dairy heifer rearing practices from birth to weaning and their associated costs on UK dairy farms. *Open Journal of Animal Sciences* 5.

2. Blowey, R. W. (1999) *A Veterinary Book for Dairy Farmers* (3rd Edition). Farming Press.

3. Heinrichs, A. J. and Jones, C. M. (2003) Feeding the newborn dairy calf. Penn State Extension Paper.

4. Charlton, S. J. (2009) *Calf Rearing Guide*. Context Publications.

5. Roberts, J. (2009) Establishing gut microflora. *Veterinary Times* 24.

6. Drackley, J. K. (2008) Calf nutrition from birth to breeding. *Veterinary Clinics of North America, Food Animal Practice* Vol. 24 No. 1.

7. National Research Council. (2001) *Nutrient Requirements of Dairy Cattle,* 7th Edition. Washington DC: National Academy Press.

8. Bartlett, K. S., McKeith, F. K., VandeHaar, M. J., Dahl, G. E. and Drackley, J. K. (2006) Growth and body composition of dairy calves fed milk replacers containing different amounts of protein at two feeding rates. *Journal of Animal Science* 84.

9. Schrama, J. W., Arieli, A., van der Hel, W. and Verstegen, M. W. (1993) Evidence of increasing thermal requirement in young, unadapted calves during 6 to 11 days of age. *Journal of Animal Science* 71.

10. AHDB Dairy Factsheet (2016) Managing calves in cold weather

11. Van Amburgh, M. E. and Drackley, J. K. (2005) Current perspectives on the energy and protein requirements of the preweaned calf. In: Garnsworthy, P. C., editor. *Calf and Heifer Rearing.* Nottingham University Press.

12. Potter, T. (2015) Calf rearing – approaches to care and performance. Veterinary Times 2.

13. Langford, F. M., Weary, D.M. and Fisher, L. (2003) Antibiotic resistance in gut bacteria from dairy calves: A dose response to the level of antibiotics in milk. *Journal of Dairy Science* 86.

14. Cooper, R. and Watson, I. (2013) A guide to feeding and assessment of calf milk replacer. *Livestock* 18 No. 6.

15. Heinrichs, A. J., Wells, S. J. and Losinger, W. C. (1995) A study on the use of milk replacers for dairy calves in the United States. *Journal of Dairy Science* 78.

16. Davis, C. L. and Drackley, J. K. (1998) *The Development, Nutrition and Management of the Young Calf.* Ames (IA) Iowa State University Press.

17. BAMN Publications (2008) *A Guide to Modern Calf Milk Replacers. Types, Use and Quality*. AFIA Publications.

18. Godden, S., Feirtag, J., Green, L., Wells, S. and Fetrow, J. (2003) A Review of issues surrounding the feeding of waste milk and pasteurisation of waste milk and colostrum. *Proceedings of the Minnesota Dairy Health Conference.*

19. Foley, J. A. and Otterby, D. E. (1978) Availability, storage, treatment, composition, and feeding value of surplus colostrum: A review. *Journal of Dairy Science* 61.

20. Der Burgt, G. V. and Hepple, S. (2013) Legal position on "once a day" feeding of artificial milk to calves. *Veterinary Record* 172.

21. Appleby, M. C., Weary, D. M. and Chua, B. (2001) Performance and feeding behaviour of calves on ad libitum milk from artificial teats. *Applied Animal Behavioural Science* 74.

22. Cooke, J. S. (2013) Rearing heifer caves for optimum lifelong production. *Cattle Practice* 21.

23. Bach, A. (2012) Nourishing and managing the dam and postnatal calf for optimum lactation, reproduction, and immunity. *Journal of Animal Science* 90.

24. Bach, A. (2011) Associations between several aspects of heifer development and dairy cow survivability to second lactation. *Journal of Dairy Science* 94.

25. Pedersen, S. (2015) Water: a forgotten nutrient. *Veterinary Times* 45:21.

Feeding from weaning to calving

Veterinary Guide To Rearing Dairy Heifers

4

Feeding from weaning to calving

With the early challenge of colostrum management mastered, the change from pre-ruminant to ruminant overcome, and a successful weaning transition achieved, only 22 months of feeding the replacement heifer remain before a return on investment is seen, and she enters the milking herd.

The average cost of rearing a heifer from weaning to conception in the UK is £754.94[1]. The largest cost associated with this is feed (35.6%), followed by labour (24.7%) and bedding (8.9%) (Fig. 1). The combined feed cost including purchased and home-grown forage plus grazed grass is 42.3% of the total rearing cost.[1]

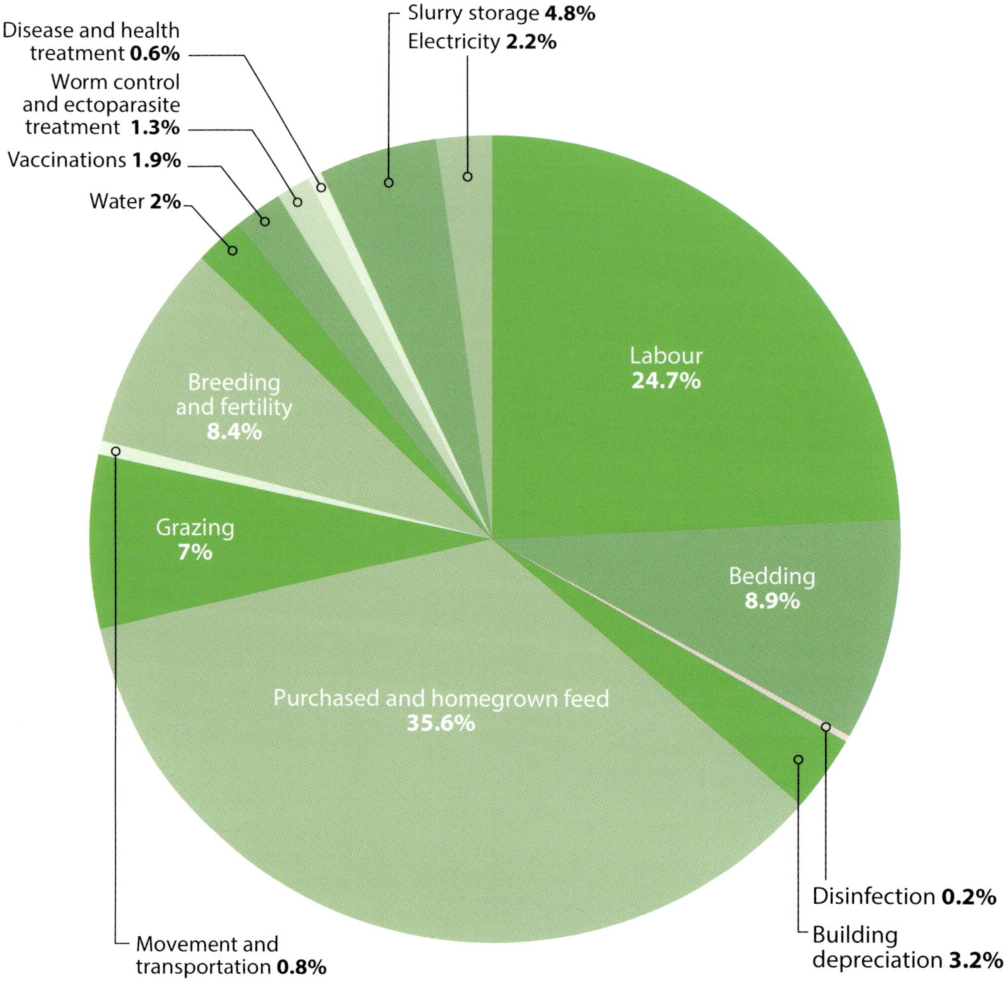

Fig. 1: Contribution of input costs to total cost of rearing from weaning to conception[1]

Feeding from weaning to calving

There are currently no figures for cost of rearing beyond conception. However, gestation length is fixed and so if the cost per day were to stay the same as before conception, the average total cost of rearing a heifer from weaning to calving in the UK could be assumed to be £1207.94.

The objectives of raising the weaned calf to calving are:

- *Optimising growth*
- *Minimising health problems*
- *Puberty at 13 months*
- *Conception by 15 months*
- *Calve at 24 months (see Chapter 6 for an explanation)*
- *Controlling costs*

This requires an understanding of the heifer's nutrient needs as well as the feed and management systems available to meet those needs.

The onset of puberty (and ability to calve) is determined by bodyweight not age and is around 40% of adult bodyweight. The period prior to puberty is the one of greatest absolute growth, with 75% of frame growth occurring.[2] To achieve age based targets, therefore, requires adequate growth rates, and bodyweight and withers height must be monitored (Figs 2 & 3).

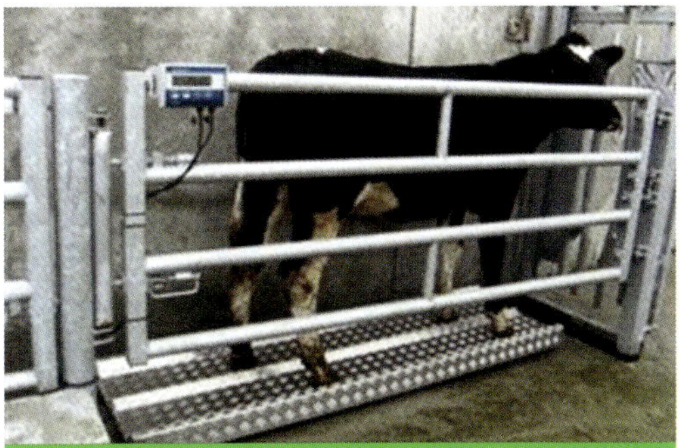

Fig. 2: Weighing a heifer on an electronic weigh scales

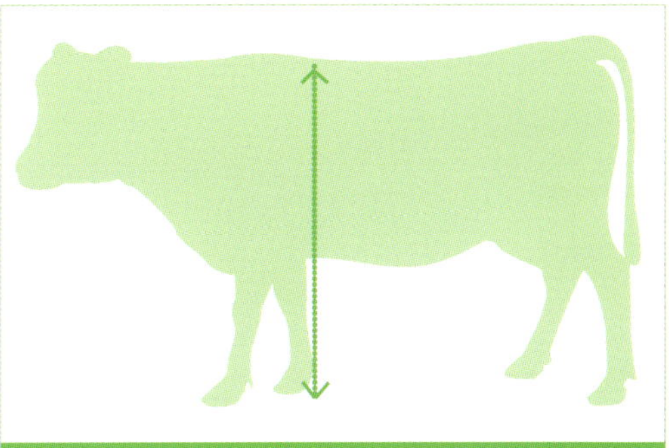

Fig. 3: Withers height

Table 1 shows suggested target body weights for Holstein-Friesians. Calculating the optimum growth rate is breed-dependant and possible for an individual herd:

- Weigh six 3rd lactation cows and average their weights (these cows are considered fully mature)
- The heifer should be 55–60% of adult body weight (about 13–14 months) at breeding. If the adult cow is 660 kg, then at breeding the heifer needs to be 360–400 kg

AGE (MONTHS)	TARGET WEIGHT (KG)	% OF MATURE BODY WEIGHT
Birth	36–40	6
1	55–60	8
3	95–110	15
6	180–200	28
9 (Puberty)	274	40
15	375	55–60
22–24 (First calving)	582	85
Mature body weight	685	100

Table 1: Suggested target body weights for Holstein-Friesians

- If calf birth weight is 40 kg then the average daily weight gain per day up to 13–14 months of age is 0.75–0.8 kg
- Withers height may be better for ease of monitoring, plus it avoids serving short fat heifers. This can be done with a yardstick while animals are in the crush or estimated by walking heifers alongside a wall with the target line clearly marked (Fig. 4).

Feeding from weaning to calving

Fig. 4: Measuring withers height by walking heifers past a mark on the wall at 129 cm = target bulling height

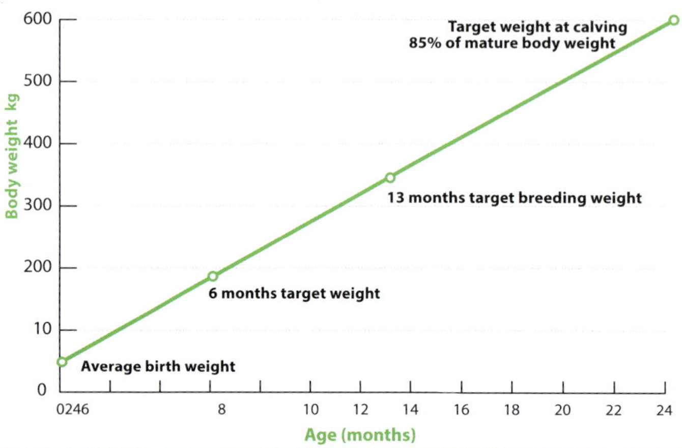

Fig. 5: What weight should the heifer be at different ages?

To work out what weight a heifer should be at different ages, a simple graph can be produced (Fig. 5). On this average birth weight for calves born is plotted and the average weight of a first calving heifer = 85% of mature cow's weight (which itself is determined by weighing six 3rd lactation cows and averaging their weights). A line is drawn between the two points and weight for each age group can be worked out by looking across to the weight axis for that specific age. Six and thirteen months are highlighted – key times in the growth of the heifer. Alternatively, weights and withers height targets for different breeds are shown in Table 2.

AGE (MONTHS)	HOLSTEIN		FRIESIAN		JERSEY	
	Weight (kg)	Withers height (cm)	Weight (kg)	Withers height (cm)	Weight (kg)	Withers height (cm)
2	76	87	72	84	55	78
3	110	93	100	88	75	82
4	127	96	120	93	95	89
6	180	104	162	100	130	94
12	340	124	285	118	220	109
15	420	129	350	122	265	114
16	440	131	370	126	280	117
18	490	133	405	130	305	119
21	545	137	470	132	355	120
22	586	138	490	133	362	121
24 (pre-calving)	636	140	535	134	395	122
24 (post-calving)	568	140	485	134	350	122

Table 2: Heifer weight and withers height targets for the main dairy breeds[3]

What about milk yields?

Growth from 3 months of age up to puberty is particularly important, as mammary development is three times faster than the body as a whole.[4] Historically, there has been concern about increased growth rates in heifers having negative effects on milk production. This is not true. Mammary fat will increase with increased growth rates and body condition. However, the secretory tissues responsible for milk are unaffected. Amounts of secretory tissue are determined by the age of the heifer and are independent of growth rates.[5, 6]

How can feed help maximise fertility?

- *Young heifers at breeding have better conception rates*
- *Since fertility increases up to the third oestrus cycle after puberty, heifers need to be grown to reach puberty at least 6 weeks before the start of serving*[3]
- *Avoid any changes to the diet composition during the 10 days before and after service*
- *To ensure successful implantation of the embryo a generous level of feeding should continue for 6 weeks after service*[3]
- *Heifers need "good" nutrition but must not get over fat*
- *"Step-feeding" with periods of restricted and then compensatory growth has little effect*

Feeding from weaning to calving

How are difficulties at calving avoided?

- *Over fat heifers are more likely to suffer from difficulties at calving*
- *After puberty there is a tendency to fatten (care with energy levels once puberty is reached)*
- *Heifers that are over-conditioned at calving will lose more weight in early lactation and be difficult to get back in calf*
- *Underfeeding heifers causes stunted growth and can delay or suppress oestrus. Breeding is thus delayed until heifers reach adequate size. This delays entry into the milking herd, reducing cash flow*
- *Poorly developed heifers are also prone to more calving difficulties than larger heifers*
- *Holstein heifers should calve at condition score 2.5-3.0 so they are not over fat but are carrying sufficient fat to mobilise early in lactation (1 kg of body fat + protein = 10 litres of milk)*
- *Target growth rates must be achieved by gaining lean body mass*

How do we achieve these goals?

1. *To calve at 24 months we need:*

   • *Early onset of puberty, achieved by appropriate growth*

   • *Optimal fertility at 13–15 months, achieved by appropriate growth*

   • *Minimal risk of difficult calving by having appropriate body condition and body size*

   ---->Daily live weight gain of 0.75–0.8 kg

2. *To control cost:*

   • *Use feeds that are available*

   • *Avoid over feeding/over fattening*

   • *Know that periods of reduced and then compensatory growth have little effect*

   ----> Monitor daily live weight gain and body condition score and adjust feeding accordingly

Feeding systems

There are several possible feeding systems for heifers from weaning to calving:

1. ***Grazing***

   • *With out-wintering*

   • *Without out-wintering*

2. ***Inside***

   • *Straw-based*

   • *Silage-based*

   • *Hay-based*

3. ***Precision feeding***

Almost any system can work. Focus must be on:

- *Protein and Energy balance* (Table. 3)
- *Set the diet to meet the nutritional requirements and promote frame growth*
- *Know the composition of feeds, especially forages*
- *Provide adequate fibre*
- *Have adequate mineral levels, particularly selenium and iodine*
- *Plan grazing and supplements to meet targets*
- *Monitor growth rates*

AGE (MONTHS)	WITHERS WEIGHT (KG)	TARGET WEIGHT (KG)	METABOLISABLE ENERGY (MJ/KG DM)	CRUDE PROTEIN (G/KG DM)
2	87	0.80	12.2	184
4	96	0.90	11.7	176
6	104	0.90	10.9	164
12	124	0.86	9.9	126
16	131	0.82	9.9	126
22	138	0.80	9.9	126
24 (pre-calving)	141	–	11.0	148
24 (post-calving)	141	–	12.0	180

Table 3: Holstein heifer nutritional requirements[3]

What and when to feed

- Immediately post-weaning:
  - There is a high risk of a growth check
  - A need to compensate for a lack of milk
  - Ad lib concentrates are required, e.g. 12.5 MJ ME/kg DM, 18% Crude Protein (CP)
- 12–16 weeks
  - Concentrate can be reduced depending on forage, e.g. 3–4 kg, 16% CP concentrate with ad-lib straw or 2 kg, 16% CP concentrate with ad-lib grass silage
  - Do not feed forages that would be unacceptable to cows
  - If not growing fast enough – feed more
- Crude protein requirements drop significantly at 12 months (Table 3). At weaning crude protein requirements are 18–19%, at service they are 12–13%
- Energy density requirements drop from approximately 12 to 10 MJ/kg DM between weaning and service (Table 3)
- If breeding at 13 months change the ration at 12 months so there are not any negative effects on fertility
- Once pregnancy is established:
  - Weights should be recorded and compared with target
  - Group heifers according to size
  - Required growth rates should be calculated using the target calving weight and feed adjusted accordingly, with higher levels of supplementary concentrate feed offered to smaller heifers.
  - High quality forages without further supplementation should be adequate for well-grown heifers[7]
- Pre-calving:
  - Two months before calving house or graze heifers with dry cows to reduce stress of herd introduction
  - If heifers are to be managed and milked as a separate group during their first lactation, pre-calving heifers can be managed as a separate group[7]

How to feed it

- *Ensure adequate feed access (Table 9, Chapter 5, page 119)*
- *Coarse mix vs pellets*
  - *Intakes may be lower with pellets, but they have a better feed conversion efficiency*
  - *Feed what works*
- *Not too much coarse fibre early on*
  - *Expecting high intakes of straw at a young age is too optimistic.*

Grouping of heifers

With smaller dairy farms grouping of heifers can become extremely difficult. A wide range of ages makes ration formulation challenging. The best approach is to always formulate the ration for the youngest animal to get maximum growth rate. At the same time monitor older animals in the group to make sure they do not become over-conditioned.

Fig. 6: Young Montbeliarde heifers turned out at grass

Fig. 7: Measuring sward height with a plate meter

Fig. 8: Topping swards to prevent the build up of low digestibility herbage that would normally occur with longer swards

Grazing

The most common criteria for turnout of heifers in the UK are grass condition (46.5%), weather (38.6%) and age of heifer (27.7%)[1] (Fig. 6). The age that heifers are turned out at for the first time and the amount of time spent at grass from weaning to conception is strongly influenced by calving pattern. The overall average age at turnout in the UK is 189 days, this varies from 129 days (spring calving), 178 days (all-year-round), 195 days (Autumn Calving) and 164 or 238 days (multi block calving).[1] The average cost per heifer per day spent grazing is £0.25.[1]

Preparing for turnout

1. Parasites
   The immune system of young heifers takes time to develop. Until it has, they are susceptible to parasites that will seriously impact on their growth rates. These risks can be controlled by, clean pasture, wormers and vaccination (see Chapter 7).
2. Diet
   Turnout can lead to a growth check, especially with young heifers. Reduce this by continuing to feed concentrates (up to 2 kg) after turnout for at least 2 weeks, to ease the transition.[8]
3. Other
   - Freeze brand bulling heifers before turnout
   - Check if any ear tags have been lost – and replace
   - Turnout into a small field to minimise charging about and injury
   - Ensure the field is well fenced
   - Make sure water troughs are turned on[8]

Turnout

Within the UK, autumn-born calves are generally at grass from 6–13 months of age, while spring-born calves are at grass from 3-10 months of age. However, there is a wide variation in age within each of these groups and grazing regimes must be adapted accordingly. Both sward and animal factors affect performance at grass.[9]

Sward factors:

- *Digestibility*
- *Height (Fig. 7) – Critical height" varies depending on set-stocking or rotational systems*
- *Density*
- *Grass management (Fig. 8) – Management throughout the season will affect quality*

Animal factors:

- *Growth potential – Breed dependant*
- *Stage of development – Calves vs heifers, pregnant vs non-pregnant (Table 4)*
- *Stocking rate – Higher levels affect live weight gains*

AGE		INITIAL LIVE WEIGHT (KG)	LIVE-WEIGHT GAIN (KG/DAY)
Calves		> 181	0.84
		136 – 181	0.79
		91 – 136	0.75
		< 91	0.66
Heifers	Pregnant	> 408	0.88
	Pregnant	363 – 408	0.91
	Non-pregnant	318 – 363	0.81
	Non-pregnant	< 318	0.78

Table 4: Size of animal at turnout and live weight gain at grass[10]

Rotational grazing systems, where cattle are moved from paddock to paddock throughout the grazing season, produce superior live weight gains relative to set stocking grazing systems, where cattle have access to the entire grazing area for the majority of the grazing season. Within rotational grazing systems, performance of younger calves is maximised if they are allowed to graze ahead of older cattle (leader/follower system) compared to grazing the two groups completely separately (Table 5). In the leader/follower system, calves select the higher-quality herbage (higher in crude protein and lower in crude fibre) and then the older cattle consume the remaining lower-quality herbage with no detrimental effects on performance.[11] If rotational stocking, then put young heifers onto fresh leafy swards and remove when 7.5 cm of grass remains. Older heifers can graze more mature swards down to 6 cm.[7]

LIVE WEIGHT GAIN (KG)	LEADER / FOLLOWER	SEPARATE GRAZING
Calves 7–12 months	0.81	0.60
Heifers 19–24 months	0.78	0.84

Table 5: Effect of leader/follower vs separate on the performance of calves and heifers[12]

Set stocking and paddock systems are alternative grazing systems to leader/follower. Table 6 outlines the advantages and disadvantages of each. In set stocking systems the stocking rate: should be adjusted to match grass available and required growth rate; aiming for a sward height of 5-6cm in spring, 7 – 8 cm in mid-summer and 9 –10cm in autumn. In this "1–2–3" grazing system, the grazing area dedicated to the heifers is divided into three blocks. In spring heifers graze one area while the other two blocks are cut for silage. Later heifers graze the two silage aftermaths while the grazed block is cut for silage. At the end of the season all three blocks can be used for grazing.[7]

Feeding from weaning to calving

Once turned out, grazed grass forms an important component of a heifer rearing system. Manage grass so as to achieve high live weight gains per animal and efficiently utilise the grass available. The main problem in the development of grazing systems for dairy heifers is that calves and older cattle have very different grazing habits.[11] As calves are more selective in their grazing habits than older cattle and graze for fewer hours per day, they are less willing to harvest a large proportion of the herbage available. As a result live weight gain can be reduced if they are forced to utilise more of the grass in a paddock before being moved to a new paddock.[10]

Fig. 9: Round bale feeder with grass silage provided as a supplementary feed

GRAZING OPTIONS	Advantages	Disadvantages
Set stocking Keep the heifers on one patch of land all season	• Does not require frequent moving of stock • Less fencing and water troughs	• Difficult to manage grass growth • Poor growth rates can easily occur if grass growth slows
Paddock system Rotate the heifers through a series of paddocks, keeping good grass in front of them	• Easier to keep good quality grass in front of young heifers • Paddocks can be taken out of the grazing round for silage	• Higher set up costs • More movement of heifers
Leader/follower system A paddock system where young heifers graze the paddock in front of older heifers	• Younger animals get the best grass with the lowest parasite challenge • Older heifers ensure the paddock is well grazed and parasites are cleaned up	• Requires lots of paddocks ideally of similar size • Frequent stock movements • Depending on fences, running a bull with the heifers can be difficult

Table 6: A comparison of different grazing options for heifers[8]

Fig. 10: Heifers out-wintered

Supplementation at pasture

Supplementation at pasture can be in the form of concentrates or conserved forage (Fig 9). The main aim of supplementation at pasture for heifers is to improve daily live-weight gain, or to maintain growth rates if grass supply is indequate.[9] However, responses to concentrate supplementation are extremely variable and depend on many factors, such as growth potential of the animal, sward height, sward quality, supplement type, composition and substitution rate. The largest responses are gained under restricted grazing conditions (sward height 7cm or less).[9]

Out-wintering

Providing they are meeting their target growth rates heifers can safely be kept outside over their first winter on block grazing or on forage crops (Fig. 10). In-calf heifers can also be kept outside, over their second winter. Out-wintering requires free-draining land, and the animals may need to be moved regularly to ensure they do not poach the fields badly. If heifers are out-wintered on grass, then offer hay or big bale silage; if they are on a "catch crop" such as stubble turnips then offer plenty of straw.

Managing fences, water and shelter

For ease of management there should be an additional electric wire at a lower level on field and track fences to prevent calves escaping. It is important to ensure calves can reach into water troughs and are not at risk of falling in. With small calves place a temporary smaller trough in their fields and move it with them. Provide shelter for smaller calves in the field. If there are no good hedges or convenient buildings available, construct shelters from big bales covered with sheeting.[13]

Inside

When inside during the rearing period, heifer diets can include a range of forage and non-forage ingredients, common examples are shown in Table 7.

	3 months	7–8 months	12 months	16 months	20 months
Weight change (kg/day)	0.9	0.9	0.9	0.9	0.8
Bodyweight (kg)	100	200	300	400	500
Grass silage-based (Kg fresh)					
11ME/12CP grass silage		10.00	22.00	28.00	32.00
Barley straw	1.00				
18% Rearing cake	3.00				
Wheat (rolled)		1.25			
Soya (hi pro)		0.50			
Total intake	4.0	11.8	22.0	28.0	32.0
Nutrient supplies					
Energy required (MJ)	35	52	72	92	104
Diet energy (% of req.)	109	104	101	100	102
Protein req. (g)	335	377	416	457	480
Diet protein (% of req.)	104	103	102	117	124
Straw-based (Kg fresh)					
Barley straw	1.00	4.00	6.50	9.00	11.00
18% Rearing cake	3.00				
18% Dairy cake		3.00	3.50	4.00	4.00
Total intake	4.0	7.0	10.0	13.0	15.0
Nutrient supplies					
Energy required (MJ)	35	57	78	98	112
Diet Energy (% of req.)	109	104	104	104	103
Protein req. (g)	335	377	416	457	480
Diet protein (% of req.)	104	106	120	131	131

Table 7: Typical Holstein heifer diets[3] Note: On a grass-based system less concentrate and additional protein can be fed. Whereas, on a straw-based system, more concentrate and bought-in protein is needed as straw has less protein.

The forage should be analysed to assess nutrient content. If you don't know the content of forages, you cannot know what you are feeding the heifer, and how she can be expected to perform. Repeat the process at least twice through the indoor period – forage quality changes over time and between bales, clamps, etc.

Alternative indoor feeds include:[14]

1. Straw and molasses
   - A good source of sugar
   - Need to balance the protein
   - Quality straw is essential
2. Maize gluten
   - Straw and gluten diets work well
   - Availability of gluten can vary
3. Straw and pot ale syrup
   - High protein and energy feed
   - Quality straw is essential

Fig. 11: Heifers feeding from a self-feed silage system

Fig. 12: Heifers feeding from an easy-feed system

Fig. 13: Heifers feeding from a TMR

Feeding systems inside

1. Self feed silage (Fig. 11)

For self feed to work well, it is vital the silage is well managed:[14]

- Roll back the sheet a little at a time and keep the front edge weighted
- Be careful not to limit feed intakes by moving the barrier too slowly – this is especially harsh on the smaller animals who will not be able to reach the silage for the majority of the time
- Very dense silage is hard on developing heifers' teeth

2. Easy feed (Fig. 12)

Dropping silage into ring feeders or troughs is an easy way of getting silage to heifers[14]

- Ensure troughs are cleaned regularly
- Forcing heifers to clean up very poor quality feed will suppress growth rates
- Silage removed from milking cow troughs can be used in heifers aged 5 – 8 months and pregnant heifers
- Keep a note of silage fed and waste removed to help determine intakes

3. TMR (Total Mixed Ration) (Fig. 13)

This allows full control and monitoring of feed intake:[14]

- Ensure feed barriers are well designed for youngstock
- Keep troughs clean
- Check mineral supply, especially if feeding straights or maize silage

Precision feeding

This system is uncommon in the UK. Precision feeding uses highly digestible feed sources in a controlled feeding environment with an emphasis on feed efficiency, reducing daily ration costs and minimising manure output.

In precision feeding, no additional free choice forages are fed, and the balanced diet is fed in the form of a TMR or mixture of forage and grain, fed once daily. Protein is fed at 14–15% CP for pre-pubertal heifers and 13–14% CP for post-pubertal heifers. Energy is balanced by formulating the diet at a fixed energy content to specifically meet the heifers' requirement.[15]

Most dairy heifer precision-feeding systems utilise corn and soybean meal as concentrate sources to provide energy and protein. As the concentrate proportion of the ration increases, there is greater opportunity for including cost-effective byproduct ingredients. Using high levels of maize silage in precision-fed heifer rations is possible, but requires careful monitoring of the heifers because a considerable proportion of the maize silage is grain.[15]

Precision feeding requires heifers to be weighed to determine the amounts of feed required to maintain growth rates. Beyond 4 months of age, heifers should be housed with other heifers as close to the same age as possible and always in groups with less than 90 kg of weight variation. This means having groups with 2–4 months of age variation at the most. Post-breeding, this number can be increased to 136 kg weight spread.[15]

Heifers will need 36–60 cm of feed bunk space each from 4 months of age to pre-calving or 22 months of age. Precision-fed heifers will not have access to feed at all times of day, thus all heifers in a pen must have access to the feed bunk at the same time. Aggressive and timid heifers are very susceptible to over or under nutrition when feed space is limited, so all heifers must be able to eat at one time. If feed bunk space is limited strategies include:

- *Group animals with peers having similar body weight*
- *Use headlocks or closely placed divider posts*
- *Feed twice daily at close intervals, e.g. feed two-thirds of the daily allotment at 7 a.m. and the remaining third at 9 a.m.; in this way the larger animals can eat more freely at the early feeding and the more timid animals at the second feeding*

Avoid straw and wood shavings as bedding; because if heifers are allowed to consume bedding the intended balance of the precision-fed diet is compromised.

At the start of a precision-feeding protocol, heifers will vocalise immediately prior to feeding, with frequency and magnitude increasing toward the next feeding. This behaviour will stop within 14 days. This is due to a moderate reduction in rumen and gut size needed to accommodate less feed, which is one of the reasons for the improved efficiency in precision-fed heifers. The transition to precision feeding requires time and commitment, similar to the time it takes to increase gut capacity after calving. As long as the heifers are growing according to the ADG goals of an operation and receiving a correctly balanced ration, they are adequately fed.[15]

An appropriate starting point is to limit the intake of 50% forage, 50% concentrate diet. This diet offers some of the advantages of precision feeding a high concentrate diet and will increase feed efficiency. Precision feeding should be discontinued and heifers adapted to normal pre-calving diets 30–45 days before calving[15].

Feeding higher-concentrate rations in a restricted manner to growing dairy heifers from 4–22 months of age leads to similar growth performance as other feeding systems. Provided the level of intake is restricted to allow for an optimal level of ADG, high-concentrate, precision-fed rations can be fed to dairy heifers successfully. This has the potential to reduce feed costs, reduce manure output, and increase feed efficiency without compromising future lactation performance.[15]

Mineral supplementation

The economically important trace elements in the UK are copper, selenium, cobalt, iodine, zinc and manganese. Trace-element deficiencies can have an impact on heifers, but the common symptoms, such as ill thrift and reduced growth rates, can be due to more common issues such as parasitism or insufficient energy in the diet.

When considering trace-element management local knowledge is important because the variable geology and soil types across the UK can lead to variable trace-element availability. This variability has the most impact on systems based around grazing or conserved forage.[16]

Sandy and free-draining soils often contain fewer trace elements than clay soils and those with poor drainage. However, there can be huge variation and the impact of grassland management (slurry application, liming, and fertiliser use) must not be underestimated. Soil testing is therefore useful in identifying gross deficiencies, but must be used in combination with forage analysis and testing of animals before deciding on supplementation.[16]

Copper

Copper is an essential trace-element involved in a number of enzyme pathways: copper-dependent enzymes are involved in multiple functions, such as energy metabolism, coat pigmentation, immunity, iron metabolism, bone growth and development.

In calves, copper deficiency can be seen at a few weeks old, but it is more common when they are 3 or 4 months of age.[17] Problems often occur during grazing in the Spring or Summer when mineral content tends to be lower. The signs that copper deficiency is an issue are variable because of its many roles. Most common are in coat colour changes (especially around the eyes and ear margins), ill thrift, anaemia and, potentially, infertility. In calves, widening of the growth plates of the long bones of the legs leads to lameness and poor growth. Diarrhoea may be seen in areas where molybdenum is present in higher concentrations.[16]

Copper-deficient pastures can lead to a primary deficiency, but secondary copper deficiency is more common, because of antagonism (blocking of uptake) by sulphur, iron and, especially, molybdenum from ingested soil.[16] Several grazing animals showing symptoms are an indication of copper deficiency.[17] Diagnosis is made from copper analysis of blood samples or liver tissue. Liver biopsies are possible but samples are more frequently collected from animals at slaughter or a post-mortem examination.[16]

Before providing supplementary copper, the true status of the farm and the animals must be determined, as over-provision can lead to toxicity. Cattle fed manufactured feed rarely suffer copper deficiency and are only at risk of copper toxicity because levels of copper in concentrate feeds are usually high relative to their requirement, and absorption is generally greater than from forage-based rations.[16]

Copper toxicity in dairy calves can be caused by extremely high levels of copper fed over a short period, but more often by moderate levels fed over a longer term. Excess dietary copper is stored in the liver and can ultimately cause liver degeneration, with the release of stored copper from the liver into the blood. This liver degeneration may initially be mild but can, ultimately, end in death.[16]

Selenium

Selenium is important for immune function, and protects tissues against oxidation and the breakdown of cell membranes (skeletal, cardiac and respiratory muscles are most susceptible). The action of selenium is linked with vitamin E, so diets low in vitamin E require higher levels of selenium and vice versa.

Selenium deficiency is most commonly seen as white muscle disease, although ill thrift and infertility may also be seen. Soils derived from granite or pumice are deficient.[17] Immediately after birth, the congenital form of selenium and vitamin E deficiency results in stillbirths or weakly calves unable to suck unaided. The delayed form of white muscle disease is most frequently seen in calves between 1 and 4 months of age, with signs brought on by sudden unaccustomed exercise, e.g. turnout to pasture.[16]

Skeletal, respiratory and cardiac muscles are most susceptible to the impact of deficiency giving rise to a number of presentations. Animals where respiratory muscles are affected will show breathing difficulties. If cardiac muscle is affected, the animal may just be found dead with no previous signs of disease. Where skeletal muscle is affected, animals may present with stiffness, abnormal gait or a sudden inability to stand, in the absence of any other symptoms. Calves with selenium deficiency are also more susceptible to infectious disease, such as pneumonia, due to delayed responses of white blood cells.[17]

Diagnosis of selenium deficiency is made based on blood sampling for glutathione peroxidase, or the identification of muscle changes at post-mortem.[16] Treatment requires injection of sodium selenate or selenite. With congenital deficiency, additional selenium and vitamin E must be provided during the last 2 months of pregnancy.[17] Selenium can cross the placenta, and both selenium and vitamin E are concentrated in colostrum, therefore, supplementation of the cow's diet will ensure a good supply to the newborn calf.

Iodine

Iodine is required for the synthesis of thyroid hormones – 80% of the body's iodine is found in the thyroid gland. Animals with iodine deficiency present with goitre (an enlarged thyroid gland in the neck), with relative thyroid size used as a diagnosis of the condition at post-mortem.

Iodine deficiency presents as late abortions and stillbirths, weak calves and increased newborn deaths, plus an increased incidence of retained foetal membranes. As with copper, iodine deficiency can be primary, resulting from iodine deficient soils or, secondary, caused by the ingestion of brassicas, high calcium ingestion, heavy bacterial contamination of feed or water, or a low level intake of feed containing linseed meal.[17]

Diagnosis of iodine deficiency relies on blood sampling dry cows to assess iodine levels, or post-mortems of stillbirths where the thyroid gland is weighed to determine if it is abnormally large. Iodine supplementation can be achieved by oral dosing 1 ml of tincture of iodine per cow per day in the water or 800 mg potassium iodide per cow every 2 weeks, by drench or added to the drinking water. Alternatively, apply tincture of iodine to the skin of dry cows – 7 ml of a 5% tincture of iodine weekly.[17] Longer-term solutions include boluses and the addition of iodine to the diet.

Cobalt

Cobalt is essential for ruminants as it is incorporated into vitamin B_{12} by rumen micro-organisms, so ruminants require a consistent supply. Vitamin B_{12} is secreted in milk and present in milk replacer, so deficiency is not seen in youngstock until they are weaned and become true ruminants[16]. Symptoms of cobalt and vitamin B_{12} deficiencies include loss of appetite, loss of body condition, muscle wastage, a rough, discoloured hair coat, stumbling gait and anaemia. Diagnosis is based on response to vitamin B_{12} supplementation injected at 2–6 mg/50 kg liveweight[17]. Long-term prevention is through the use of boluses.

Zinc

Zinc is involved in many biochemical processes in the body. Poor growth rates, altered gait and stiff joints, dry, scaly skin, failure of wound healing, and reduced reproductive performance are all signs of deficiency. As a supplement, organic zinc can be used to improve foot hardness and reduce incidence of sole abscesses.[16] A consistent supply of dietary zinc is required as body stores are not easily mobilised.

Manganese

Manganese regulates the activity of certain enzymes. Deficiency results in poor growth rates, impaired skeletal development, stiff gaits and swollen joints. Manganese deficiency can also result in calf deformities, including jaw abnormalities and dwarfism.[16]

How to supplement minerals

Upon diagnosis of a deficiency, various options for supplementation exist. Choice depends on minerals needed, cost, availability and ease.

Fig. 14: Free-choice mineral lick for mineral supplementation

Fig. 15: Mineral supplementation with a ruminal bolus

The easiest way of providing supplementation is free choice minerals in licks or blocks (Fig. 14). The difficulty however, is the variability in dosing, with some animals getting little or no supplement, while others take in more than required.

Compound feeds provide an easy way of supplementing a heifer's mineral intake, by adding minerals to concentrates. Inclusion rates must take into account planned feeding regimes to ensure required mineral supplementation is achieved, while avoiding potential toxicity. An alternative is diet formulation, where choice of constituents is chosen to make sure mineral deficiencies are addressed by inclusion of specific feed types.

Direct administration of required minerals to animals via injections, oral drenches or ruminal boluses (Fig. 15), has the advantage of ensuring each animal receives the prescribed dose of minerals and trace elements. However, it is more labour-intensive, involving handling of the animals, as well as having a potentially a higher cost. If chosen it is important dosing regimens take into account the duration of action, as some products may require repeated dosing.

The most important thing to consider before feeding vitamin/mineral supplements is the vitamin and mineral contribution of all inputs already in the diet. The level of mineral and vitamin supplementation required by an animal depends on its stage of development but also the forage composition of the diet and availability of minerals and vitamins from the raw constituents.

Over-supplementing minerals and vitamins to youngstock, in an effort to assure dietary adequacy, is not uncommon. It increases costs and can sometimes lead to toxicity. To ensure proper levels of minerals and vitamins are fed, test forages and feeds for their mineral content; and then provide supplements to reach requirements. Avoid free-choice mineral and vitamin supplementation to minimise the potential for variation in intakes.

Water

Water remains as important post-weaning as it is pre-weaning. Continue to offer plenty of clean fresh water to heifers, and monitor quality every 6 months (see Chapter 3, page 72).

Fig. 16: Low protein diets. (Reproduced with permission from *Diseases of Cattle in Australasia*, published by VetLearn, Wellington, New Zealand)

What can go wrong?

- Energy and protein balance
  - Avoid over-fattening – it must be lean growth
  - More protein needs to be fed if heifers are getting fat pre-puberty
  - Less energy needs to be fed if heifers are getting fat post-puberty
  - A low protein diet is the most common factor affecting growth rates and disease incidence. Heifers are small, in poor body condition and "paunchy" with distended abdomens full of poor-quality forages. Heads often appear larger than expected (Fig. 16)
- Overcrowding
  - Lack of feed space can drastically reduce feeding times in heifers
  - At 200% overcrowding (20 heifers for 10 feed spaces), time spent eating feed will reduce by 40 minutes per day
  - Average weight and total feed consumed will appear normal. However, there will be huge variation in body weight in overcrowded pens due to bullying and feed sorting
  - This can lead to breeding challenges and chronic foot problems in heifers that have to repeatedly over-reach for feed and splay their front feet, permanently altering their stance
  - Providing adequate feed space will allow all heifers to feed together, consume the same diet and minimise hoof problems (Fig. 17)
- Trace elements
  - Deficiencies can arise if no concentrates are fed
  - Pregnant heifers should only need good quality forage and a mineral and vitamin supplement
- Disease
  - Coccidiosis
  - Pneumonia
  - Parasites – gut worms, fluke, lungworm
  - All have massive impacts on growth rates and productivity and in many ways are much more significant than diet (see Chapter 7)

Fig. 17: Adequate feed space to allow all heifers to feed at the same time

References

1. Boulton, A. C., Rushton, J. and Wathes, D. C. (2015) The management and associated costs of rearing heifers on UK dairy farms from weaning to conception. *Open Journal of Animal Sciences* 5.
2. Owens, F. N., Dubeski, P. and Hanson, C. F. (1993) Factors that alter the growth and development of ruminants. *Journal of Animal Science* 71.
3. DairyCo (2012) Managing youngstock feeding. Section 10, *Feeding+, Feeding Improvement Programme.*
4. Desjardins, C. and Hafs, H. D. (1969) Maturation of bovine female genitalia from birth through uberty. *Journal of Animal Science* 28.
5. Meyer, M. J., Capuco, A. V., Ross, D. A., Lintault, L. M. and Van Amburgh M. E. (2006) Developmental and nutritional regulation of the prepubertal heifer mammary gland: I. Parenchyma and fat pad mass and composition. *Journal of Dairy Science* 89.
6. Meyer, M. J., Capuco, A. V., Ross, D. A., Lintault, L. M. and Van Amburgh M. E. (2006) Developmental and nutritional regulation of the prepubertal bovine mammary gland: II. Epithelial cell proliferation, parenchymal accretion rate, and allometric growth. *Journal of Dairy Science* 89.
7. Drew, B. (2004) Heifer rearing – 12 weeks to calving. In: Andrews, A. H., editor. Bovine Medicine, 2nd Edition, Blackwell Publishing.
8. Kingshay Farming Trust. (2006) Heifers at turnout. Farming notes from Kingshay.
9. Dawson, L. E. R., and Carson, A. F. (2005) Grazing systems for dairy herd replacements. In: Garnsworthy, P. C. (Editor) *Calf and Heifer Rearing.* Nottingham University Press.
10. Leaver, J. D. (1974) Grazing management for dairy calves and heifers. *Irish Grassland Animal Production Association Journal* 9.
11. Dawson, L. E. R., and Carson, A. F. (2004) Management of the dairy heifer. *Cattle Practice* 12.
12. Leaver, J. D. (1970) A comparison of grazing systems for dairy herd replacements. *Journal of Agricultural Science* 75.
13. DairyCo (2012) Rearing heifers at grass. Factsheet 14, Grass+, Improvement through grassland management.
14. Kingshay Farming Trust. (2006) Feeding heifers this winter. Farming notes from Kingshay.
15. Zanton, G. and Heinrichs, A. J. (2008) Precision feeding dairy heifers: Strategies and recommendations. Penn State Extension Paper.
16. Potter, T. (2014) Importance of youngstock mineral supplementation. *Veterinary Times* 44:17.
17. Andrews, A. H. (2004) Other calf problems. In: Andrews, A. H. (Editor). *Bovine Medicine* (2nd Edition). Blackwell Publishing.

Housing

Veterinary Guide To Rearing Dairy Heifers

5

Housing

Housing systems must provide:

- ***A comfortable, well-drained lying area***
- ***Shelter from adverse weather***
- ***Space to allow the animal to move, lie down and rise freely without risk of injury***
- ***Access to adequate food and water to maintain health and vigour*** (Figs. 1 & 2).[1]

Fig 1: An individual calf pen in Zimbabwe. The calf has a comfortable well-drained bed, shelter from the sun and rain, space to move, lie-down and stand, and fresh food and water

Fig. 2: Group pens in the UK. There is adequate space for these calves, but it is an uncomfortable environment with poor shelter and no food and water. This housing is not only inadequate, but represents a welfare issue for the calves.

When considering housing the first question to ask is how many calves and heifers need to be housed? Sizing of heifer replacement facilities is dependent on many herd-specific factors including:

1. *Size of the adult cow herd*
2. *Replacement rate of the herd*
3. *Calving interval of the herd*
4. *Ratio of female to male calves and use of sexed semen*
5. *Mortality rates at different stages of development*
6. *Age at which male calves are removed from the herd*
7. *Age at first calving*[2]

Calculations should be performed to determine expected numbers, but on average the number of replacements is approximately 90% of the adult herd.

There are many options for housing cattle pre- and post-weaning, none of which is incorrect. This is because there is more than one way to do things correctly - provided you have a system.

Housing pre-weaning

The principles of good calf housing are:

- *No contact with older animals or their environment (air, water, feed, bedding, pasture)*
- *Minimal contact between multiple calves*
- *All in – all out, with time to clean, disinfect and rest housing between calves*
- *Draught-free but well ventilated, i.e. air flow <2 m/s*
- *Bedding: clean, dry, abundant, also comfortable*
- *Easy to clean and disinfect*
- *Prevent stress/injury (calves and people)*
- *Ease of handling*

Calves are usually either housed individually in hutches or pens, housed in groups, or a combination of both. The pros and cons of each system are considered in Table 1.

	PROS	CONS
Individual housing	• Can monitor and control milk intake • Spot sick calves – evaluate appetite, demeanour, etc. • Reduce multiple calf-to-calf contact and subsequent risk of disease spread	• Labour intensive • Operator comfort (depending on housing system, season and region)
Group housing	• Labour savings • Allows use of ad-lib, automatic or computerised feeding systems • Operator comfort • Exercise and socialisation	• Competition for resources (feed) • Direct contact between several calves (nose-to-nose, cross sucking, shared nipple) > Potential for increased disease transmission > Increased morbidity / mortality

Table 1: The pros and cons of individual and group housing

Housing calves individually

The width of any individual stall or pen for a calf must be at least equal to the height of the calf at the withers as measured in the standing position. The length must be at least equal to the body length of the calf, measured from the tip of the nose to pin bone multiplied by 1.1. In practice this means pens at least 1.5 x 0.9 m, but preferably 1.8 x 1.0 m.[1] Calf pens must be large enough to allow calves to groom themselves, lie down and stretch their limbs and rise without any difficulty, and must also allow visual and tactile contact with animals in adjoining pens/hutches[1] (Fig. 3).

Fig. 3: Individual calf pens

If the calves are housed in individual pens, key features include:

1. *East-west orientation of housing to avoid extreme afternoon sun exposure in pens.*
2. *A single row of pens or two rows where the barn is limited to a width of 11 m or less.*

   – In single-row calf housing, new calves are placed in freshly cleaned pens and there is usually a space between them and the oldest calves in the barn. In barns with two rows of pens, both rows can be filled simultaneously and in parallel. In barns with 3 or more rows, it almost always results in vulnerable young calves directly across from older calves that are potentially shedding pathogens.
3. *Pens are separated from the outer wall by at least 1 m of space to avoid cold air from dropping directly into the pen from the eaves or space boarding (Fig. 4).*
4. *Solid panels between each calf and open mesh front and the rear panel can be mesh or solid to about 60 cm with mesh above. With an open front and open rear panel, there is greater opportunity for breezes to move through the pens, whereas solid panels on all sides of a calf pen can impede natural ventilation.[3]*

Fig. 4: This row of individual calf pens has a metre gap between the back of the pens and the outer wall. This prevents cold air falling into the pen from the space boarding, and causing a draught. It also has the advantage that calves can be bedded up from the back of their pens (so bedding doesn't end up in feed and water troughs at the front); also back ends of calves can be checked for scour

Calf hutches can be used to house calves in groups (Fig. 5), but usually are individual (Fig. 6). The abundance of fresh air means calves in hutches are often healthier than those kept in buildings. Each hutch must have an outside run. The hutches should be situated on free-draining concrete or a porous base, such as chalk or gravel, ensuring that any effluent goes to a suitable site for disposal [1](Fig. 7).

Fig. 5: A larger hutch for group housing

Fig. 6: Individual hutches

Fig. 7: These hutches are poorly situated – the angle of the hutch means effluent will accumulate under the bedding of the calf. Its better to position the hutch on a flat surface of chalk, gravel or concrete

Fig. 8: The back of this row of hutches is to a prevailing south-westerly wind. They are on a forward sloping area to allow urine to drain out of the hutch

The most common way to place hutches is outside in rows but they can also be placed under a roof. Place the back of the hutches towards the most common wind and rain direction, on a forward sloped area. This means urine drains out from the hutch and bedding is not exposed to rain (Fig. 8).

A good logistic flow is required for feeding, bedding and cleaning out the hutches; before moving in new calves. Hutches should be placed in rows sideways. Create enough space between the rows of hutches so they can be moved forwards or backwards, creating a new row (Fig. 9). Used hutches in the first row are moved to the second (empty) row. Between calves, the hutch should be cleaned and disinfected, and turned over to expose it to sunlight (UV) that provides and additional disinfectant procedure[4] (Fig. 10). The first row can then be easily scraped by a tractor or other equipment. Rotate back to original location for the next calf and cleaning procedure.

Fig. 10: Upturned hutch following cleaning, making use of the additional disinfectant properties of UV light

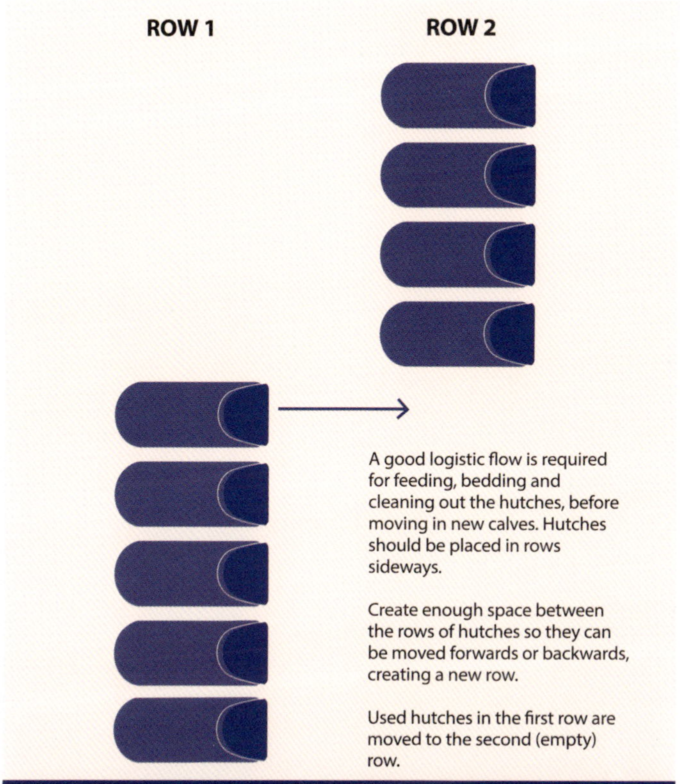

Fig. 9: Diagram demonstrating how calf hutches can be moved between rows

If the hutches are placed under a roof or in other buildings, it is important to allow and plan for two locations for each row of hutches. This reduces transmission of disease and allows the area to dry between each new calf. Never block or cover the front opening of a hutch with blankets or plastic sheets.

Calves can be kept in hutches up to 8 weeks before transferring to a group facility. Hutches are great individual housing for calves and often considered the gold standard. The major drawback is that being outside checking and feeding of calves is required in all weather conditions.

Housing calves in groups

Calves can be housed in groups from birth (Fig. 11), but from 8 weeks of age, calves must be group housed (unless an animal is kept in isolation on the advice of a veterinary surgeon).[1] Table 2 shows unobstructed space requirements for calves kept in groups. Calves should ideally not be grouped more than 12 to a pen and definitely not more than 20. If calves are grouped in numbers above 20 then stocking density is self-correcting – numbers will get down to 20 one way or another!

Fig. 11: Calves housed in groups from birth

Air space is also critical, with a minimum requirement of 6 m^3 per calf at birth, increasing to 10 m^3 by 2 months of age and at least 15 m^3 by 6-7 months.[1] The greater number of calves in a single air space, the greater the risk of disease (Fig. 12).

Weight of calf (kg)	Approximate age (months)	Minimum (statutory) area (m^2/calf)	Recommended area (m^2/calf)
45	0	1.5	2.0
46–99	0–2	1.5	3.0
100–149	3–5	1.5	4.0
150–199	5–7	2.0	5.0
>200	7+	3.0	6.0

Table 2: Unobstructed space requirements for calves kept in groups[1]

Fig. 12: Overcrowded calves, plus sharing the same air space as adult cattle = a much greater risk of disease

Housing post-weaning

Housing and feeding heifers after weaning is a challenge. This is because herd sizes in the UK often mean different-aged animals have to be kept together in the same pens or buildings.

Leave calves in their individual pens or groups for 7 days after weaning to avoid additional stresses of changes of environment and group competition. Then place individually housed calves into groups of 4–6 calves housed together in a separate pen (Fig. 13). The weaning transition is one of the two most stressful periods in the life of the young calf – so providing these "transition groups" helps to minimise growth slumps and disease risks at this time. It also allows observation for calves that may suffer disease and fail to thrive.[5]

Fig. 13: "Transition group" of 4 calves penned together 1 week after weaning

Put the transition groups together after a further 2 weeks – provided there are no health problems. Group-reared calves should be placed into larger groups 2 weeks after weaning.

House transition calves, from the post-weaning period to 6 months of age, in small groups of 10 calves or less, with a weight range less than 50 kg or an age range of no more than 2 months. This allows calves to learn social behaviour while limiting competition at the feed face. Close attention needs to be paid to transition calves' water and feed consumption as they learn to drink from a new water system and eat at a feed face, potentially with headlocks or a slat bar feed face. Ensure water troughs are low enough for small calves to easily find and drink from.[2]

Fig. 14: Heifers in cubicles

For growing heifer groups, it is also important to minimise the weight variation within a pen. This becomes more critical when cubicles are used, because inappropriately sized cubicles can injure heifers that are too big for them. Limit the weight range to about 100 kg within the group. When heifers are growing at a rate of 0.80 kg/day, the age range must be 115 days or just less than 4 months[2] (Fig. 14).

Cubicles must provide a clean, comfortable lying space, allow the heifer to enter and exit easily, and lie down and rise without interference or injury. Poorly designed cubicles and inappropriate management can lead to problems such as cubicle rejection in adult life (Fig. 15), wet and soiled cubicle beds and physical injury to the animals[1].

Fig. 15: Cubicle rejection

Weight of heifer (kg)	150	200	300	400	500
Width (m)	0.6	0.7	0.85	0.95	1.10
Length (m) (against wall)	1.6	1.7	1.95	2.15	2.4
Length (m) (head to head)	1.5	1.6	1.8	2.0	2.25

Table 3: Recommended cubicle dimensions for different sizes of heifers[6]

Fig. 16: Cubicle size can be adjusted with brisket boards and head rails

Fig. 17: Suspended cantilever cubicle divisions allow easy adjustment of cubicle height and width

Table 3 shows different dimensions for different weights of animal. Knowing the weight of heifers at each stage is therefore essential. Cubicles are only an appropriate housing option for heifers greater than 180 kg.[2] The length of the cubicle should provide body space, head space and lunging space and it needs to be adequate to allow the heifer to rest comfortably and rise without injury.

When rising from a lying position the heifer lunges forward, transferring her weight from hindquarters onto forequarters. In doing so her head is thrust forward. This lunging space must be provided for in the design of the cubicle – if it is not difficulty in rising will be experienced.[1] It is always safer to have a cubicle that is too long, as the effective length can always be reduced. Position when lying and standing can be controlled by brisket boards and head rails. A correctly located heifer means urine and dung will fall into the scraped passage.[1] (Fig. 16).

The width of the cubicle must also allow the heifer to lie and rise easily but, if excessive, might allow her to lie at an angle or turn around. It will be determined by the size of a heifer but also the choice of cubicle division, with slightly wider widths required if there is a rear support leg. There are several division designs on the market, however, the suspended cantilever offers the best option for growing heifers (Fig. 17). The main benefit is that both height and width spacing can be altered at any time – providing flexibility, especially where animals are growing. In addition, it also offers more room allowing slightly narrower widths.[1]

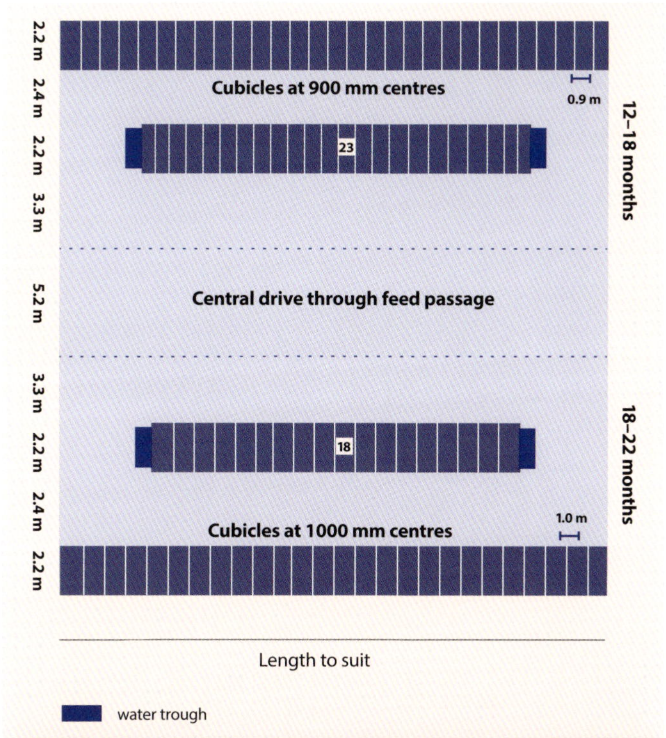

Fig. 18: Example of heifer cubicle housing – suitable to house 49 animals at 12–18 months of age and 39 animals at 18-24 months of age (5% additional cubicles), with a central drive-through feed passage.
(From AHDB Dairy Housing - a Best Practice Guide[1])

Always have 5% more cubicles than heifers within a group. Overcrowding leads to reduced lying times and increased lameness, which is often carried through to adult life. There is also more bullying with an increased risk of injuries. As with layouts of cubicle buildings for adult animals, there should be no dead ends, and cross passages provided at approximately every 20 cubicles (Fig. 18).

Straw yards provide a flexible alternative to cubicles for housing replacement heifers. The number and size of heifers housed in each group can be adjusted as needed, as long as the feed face dimensions are suitable for the heifer size.[2] The bedded area should be supplemented with a concrete feed passage, which will help promote hoof wear and prevent feet becoming overgrown.[1] Bedding must be added regularly, and the straw yard designed to allow easy removal of soiled bedding and manure. This can be done with an automatic scraper or scraper tractor, and a gate system to hold heifers back to either the feed alley or bedded area during cleaning (Fig. 19).

Fig. 19: Heifers on a straw yard

The concrete passage should be 2 m wide for heifers less than a year, and scraped at least three times per week. A small step (0.2–0.3 m) should separate feeding and bedded areas. This retains the straw and prevents manure flowing onto the bedding during scraping.[1]

Management of housing systems

There are key areas that must always be kept in mind when assessing housing, designing new housing, or making alterations to existing housing:

1. **Moisture**
2. **Fresh air**
3. **Air speed**
4. **Temperature**
5. **Light**
6. **Space requirements for feed and water**
7. **Cleanliness and hygiene**

Moisture

Moisture is ever-present in animal housing (Table 4). Bacteria and viruses that cause scour and pneumonia thrive in moist environments, and find it hard to survive in dry conditions. The aim must be to keep moisture out, and what the animals produce must be kept under control (Table 5, Figs 20 & 21).

Factor	Condition	Contribution	Symptoms
Moisture	Too much	• Supports microbial activity • Promotes bacterial growth (some species) • Absorbs energy • Acts as a transport medium • Increases slippery floors – stress	• Dirty water lying • Dirty cattle • Damp floors in areas that could be dry • Water ingress • Leaking drinkers • Condensation • Staining of underside of roof • Animal health....

Table 4: The effects of too much moisture[7]

Competent drainage	1 in 20 slopes under straw. 1 in 10 under any feeding or drinking areas. Channels in floors to intercept excess liquors. Drainage within pen, between pens, within building, outside
Competent ventilation	Heat and moisture are continually produced and must be continually removed, along with bacteria, viruses and gases
Dry bedding	To maintain thermal comfort and to absorb excess moisture at animal level Good straw and manure management
Sympathetic building materials	Cold surfaces promote condensation, keeping moisture in the building. Consider concrete and use of tin
No additional moisture	Moisture can be prevented from entering a cattle building by attention to downpipes and gutters, good building maintenance, and keeping groundwater away by sensible drainage

Table 5: Options for improved moisture management and control

Fig. 20: Standing water is a source of moisture in this building. Not only does it provide a medium for pathogens to replicate but also every time machinery or people pass through it, they can be spread around the building

Fig. 21: Broken, leaking drains are another source of moisture accumulating in a building

Moist environments are also cold environments; that lead to an increase in the rate of heat loss from a warm body, with negative effects on health and performance, particularly for calves. Table 6 shows the impact on a calf's body temperature when lying on a damp bed. The lower critical temperature (LCT) is the temperature below which the calf has to use energy to help keep it warm. A calf lying on wet straw starts to divert energy from growth to maintaining body temperature at 10°C compared to a calf lying on dry straw, which will start diverting energy at the lower temperature of 8°C.[8]

Relationship with floor	LCT °C for a 50 kg calf
Calf lying on bare concrete	17
Calf lying on dry straw	8
Calf lying on wet straw	10

Table 6: LCT for a calf lying on different surfaces (well fed and in calm air)[8]

Prolonged periods of exposure to moist conditions can also lead to physiological stress, with the eventual outcome that the calf is unable to maintain an adequate immune system. These cattle are more prone to infections.[8]

Bedding and drainage are essential to minimise accumulation of moisture. Bedding must keep calves and heifers clean and dry to minimise soiling and wet hair coats. Bedding must be comfortable for the calf to lie on – comfortable calves lie down for more than 19 hours a day. Long wheat straw is excellent and allows good insulation.[4] Deeply bedded resting surfaces are critical for very young calves in cold weather. They allow the calf to build up a layer of heat within the bed and minimise heat loss.[3]

Fig. 22: Nesting score 1
The entire rear leg is visible

Fig. 23: Nesting score 2
The rear leg is partially obscured by bedding

Fig. 24: Nesting score 3
The rear leg is completely obscured by bedding

A "Nesting Score"[9] can be used to evaluate the sufficiency of bedding based on the visibility of the rear leg of a calf lying down in the bedding. A nesting score 3 is the target for all calves and heifers (Figs. 22–24).

The floor must provide a safe durable base to the pen and surrounds that can be cleaned effectively. It should have a non-slip surface but not be abrasive to the animals' feet. To support a healthy environment drainage is very important.[8]

Health problems arise when moisture accumulates at floor level or on the bedding surface. Pooled liquid adds moisture to the air in the building, increases bedding costs, and creates a colder surface for animals to lie on.

Consider where the movement pressures are in a building. Excess water will accumulate:

- *Around water bowls and troughs*
- *Within 2–3 metres of an automatic calf feeder* (Fig. 25)
- *Along the front of single calf pens*[8] (Fig. 26)

How much is too much moisture? Answer: When you can see it!

Provide these areas with an additional slope of 1 in 10 and channels to specifically gather and remove excess liquid from the animals' environment. Calves spend up to 80% of their time lying down. A dry bed is essential to maintain health and minimise stress levels. Have a 1 in 20 gradient below bedding, plus:

- *Have drainage channels to collect and carry away excess moisture or*
- *Have a porous surface under the straw, such as slats or hardcore*
- *Regular bedding up*

Liquids moving from a pen should not move into an area exposed to foot traffic of workers. The slope of a pen can be to the front or rear with drainage channels located accordingly and designed to then carry liquids to collection points outside the building.[3]

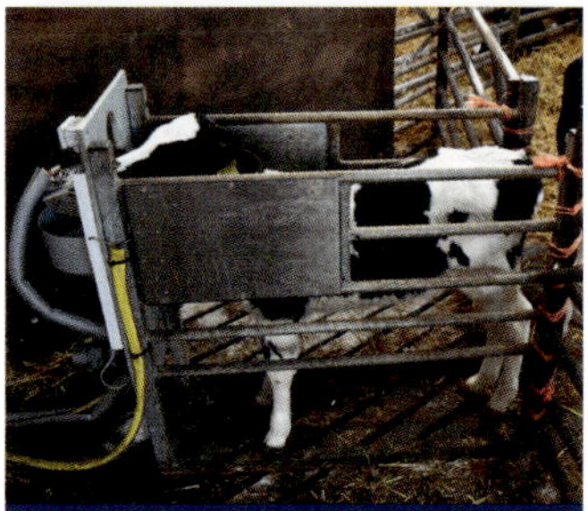

Fig. 25: The feeding station of this computer-controlled feeder is located over slats to prevent build up of moisture

Fig. 26: Moisture along the front of single calf pens

The "Kneel test" (Figs 27 & 28) provides an easy assessment of bedding:[4]

- *Soiled knees = insufficient bedding, risk of bacterial challenge*
- *Wet knees = insufficient bedding, taking heat from calf*

Fig. 27: The kneel test

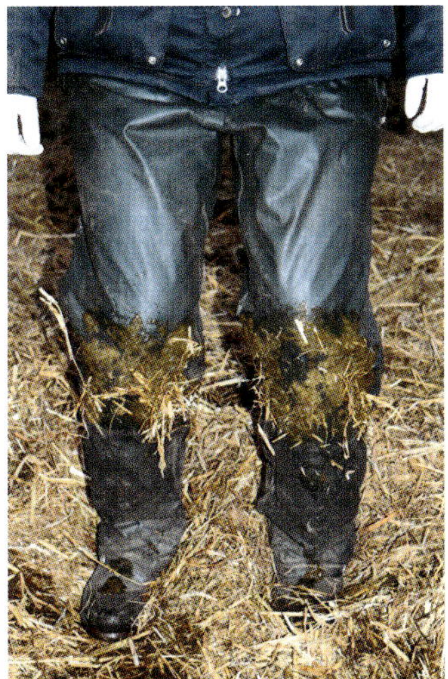

Fig. 28: The kneel test – showing soiled knees

Use the best straw on the unit for bedding calves and heifers. Damp bedding materials reduce warmth, do not absorb urine as well, and contain large concentrations of mould spores and toxins. Dusty, mouldy materials in bedding are a risk factor for pneumonia.[8] Straw choppers should not be used in buildings with stock below 3 months of age, as the normal clearance mechanisms of the lungs are not fully developed (Fig. 29). Straw choppers should also be used with caution with older stock that are showing respiratory symptoms.[8]

Fig. 29: Straw choppers are not suitable to bed up calves younger than 3 months of age

Fresh air

Providing fresh air for animals to breathe is vital. It promotes respiratory health, and maximises growth rates and feed conversion efficiency. Airborne transmission is an important route of infection for pneumonia.[8] Fresh air is free and kills bacteria and viruses: 100% fresh air kills them 10 times more quickly than 50% fresh air.[7] The way to ensure a building has adequate fresh air is to have a good ventilation rate.

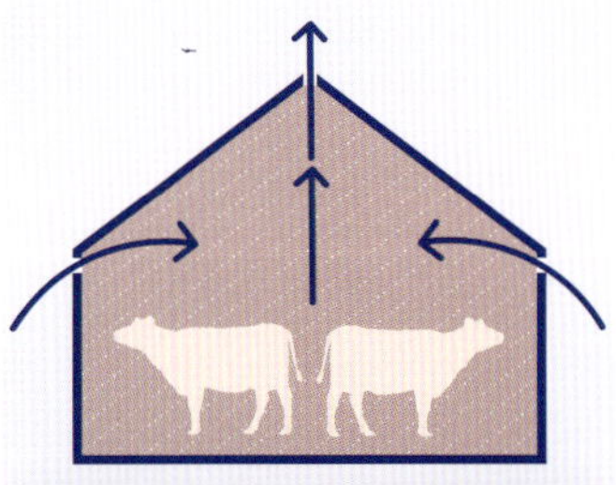

Fig. 30: The stack effect

Ventilation rate is the amount of air replaced within a building within a given time. The minimum change required within a building is 10 times each hour, increasing in the summer to 60 air changes per hour.[1] Natural ventilation works on the principles of the stack effect (Fig. 30). The heat of animals in a shed warms the air around them causing it to rise and exit through the outlet/exhaust in the roof. This then causes a negative pressure within the building that draws fresh air through the inlets.

It is important a building is designed to allow the ventilation to:[1]

- *Remove excess heat*
- *Remove excess water vapour*
- *Remove pathogens, dust and gases*
- *Provide a uniform distribution of air*
- *Provide correct air speed for stock*

The design requirements are not easy to achieve given that most buildings have no moving parts except the doors, the weather is changeable and can come from all directions, and the internal conditions including numbers and weight of animals constantly change. However, a few basic rules can be applied to assess whether the ventilation of a building is providing a healthy environment for the cattle living inside it[8]:

1. It is essential to have an outlet in the roof to let warm, moist, foul air escape
2. Outlet area needs to be 0.04 m^2 per calf and 0.1 m^2 per growing or adult animal
3. Outlets work well at ridge height, but can be spread across the roof area
4. Inlet area, ideally spread across both sidewalls, needs to be twice the outlet area (minimum), and ideally four times the outlet area
5. Inlet area must be able to control air speed, i.e. reduce wind speed
6. Gap size in space board should never exceed 25 mm
7. The outlet must be at least 1.5 m above the inlet
8. Air speed must be controlled at animal height

Well designed inlets have two principle design requirements. Firstly they allow adequate fresh air in along both sidewalls of the building, so there is no restriction to stale air exiting through the roof (Fig. 31). Secondly, inlets should reduce wind speed at animal height so that draughts and excessive heat loss does not occur. While large single inlet areas such as open gates and doors may provide sufficient fresh air to an area within a building, they are not suitable for controlling wind speed across the animals. Doors and gates must not be considered part of the ventilation system. Uncontrolled air speed at animal height is only likely to be beneficial during warm, summer months.[8]

Fig. 31: This building has adequate inlets along both sides of the shed with a continuous open-ridge providing a good outlet

The outlet area is best provided by a narrow opening for the length of the ridge, 150 mm to 350 mm wide. This is wholly dependent on numbers of animals in the shed, and building design. The wider the opening the more likely rainwater is to come in. In this case a covered open ridge is appropriate (Figs 33 & 34). A ridge like this should also be used above cubicles or anywhere rainwater entry could be a problem.[8] There are a number of methods to achieve adequate outlet ventilation, including various ridge designs or slotted roofs.

Fig. 32: There is no outlet in the roof of this calf building – clearly demonstrated by lots of cobwebs. This means there is no escape for noxious gases, dust and pathogens, which will circulate and increase the risk of disease

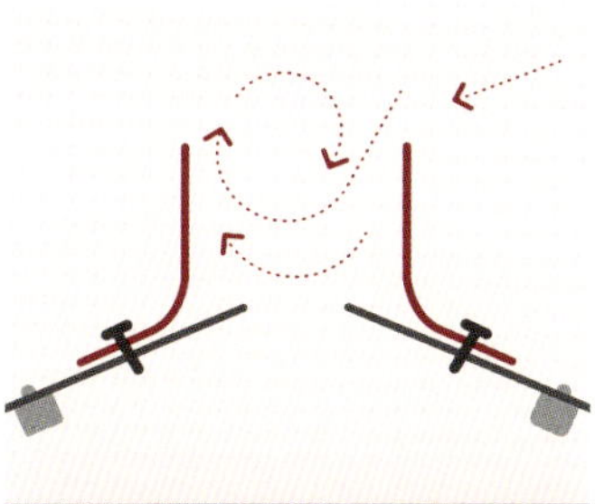

Fig. 33: A covered open ridge. When the wind moves across the ridge it circulates in the covering, this creates a negative pressure that has the advantage of drawing air out through the ridge

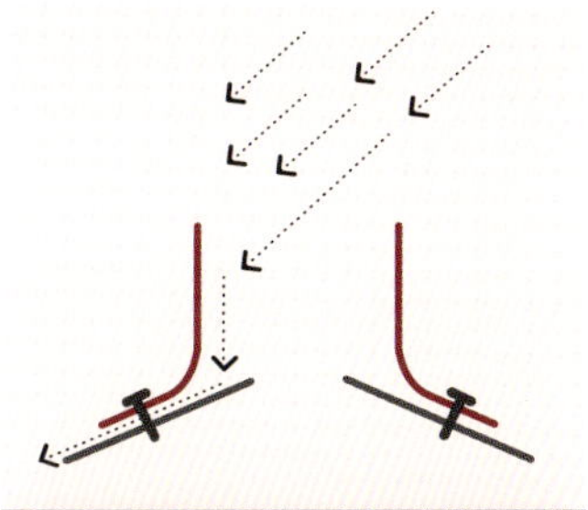

Fig. 34: A covered open ridge. When rain hits the inside of the covering it runs down and onto the roof of the building, reducing the amount of water that enters the building

In addition to inlet and outlet design a number of other points to be considered in the design of a successful natural ventilation system are:[1]

- *Buildings will naturally ventilate best when they are sited at right angles to the prevailing winds*
- *Span of the building (narrower is better for young calves)*
- *Location of the building relative to other buildings or obstructions (buildings and trees disrupt air flows for a distance of 5–10 times their height)*
- *Pitch of the roof (17–22°C is best)*
- *Stocking rate*
- *Weight of each animal*
- *Bedding system*

Fig. 35: Positive pressure ventilation tube

Naturally ventilated buildings have some shortcomings. When the wind is still, naturally ventilated buildings rely on the stack effect for ventilation. Unlike adult cows, calves do not generate sufficient heat to drive this stack effect. A further limitation of naturally ventilated buildings arises when the air outside is warmer than the air inside the building. This occurs for several hours almost every day as the sun warms the air outside more quickly than inside. During these periods, air entering the cooler interior of the building through eaves will rise and leave without good mixing near the floor.[3] In these situations the use of positive pressure ventilation systems can be used to supplement natural ventilation in calf buildings (Fig. 35).

The supplemental positive pressure tubes are usually sized to provide four changes of air per hour. This ventilation rate assumes a "normal" stocking density. The tube fan never stops, running 24 hours a day and 365 days per year. If designed properly, the tube systems will deliver fresh air without a draught into microenvironments within the calf building. The air introduced through the tubes is distributed around the building and then exits passively through the outlets.[3]

Almost all calf sheds would benefit from having a supplemental positive pressure ventilation tube, due to the inability of young calves to drive the stack effect. An evenly distributed outlet is still required in the roof, to allow the air pushed in to escape.

Air speed

Air movement for the delivery of fresh air into buildings is very important, but it must not in any way compromise health. Air speed in the part of the building where cattle stay should not exceed 0.2–0.5 m/s. If this happens the ambient air temperature may result in an increased cooling of the surface temperature of the animal (producing a draught).[6] Increased air speed at animal height increases the risk of immune suppression, and severely increases the negative impact of low temperatures on youngstock.[8] Conversely, low air speeds into buildings are associated with insufficient delivery of fresh air (Table 7).

Factor	Condition	Contribution	Symptoms
Air Speed	Too much or Too little	• Too much: associated with excessive energy loss	• Animals avoiding certain areas • Huddling • Hairy coat • High intake/low production rates • Animal health
		• Too little: associated with lack of fresh air	• Animals avoiding certain areas • Smell • Animal health...

Table 7: The effects of too much or too little air speed

Air speed is the most important factor in terms of immediate impact on youngstock. A draught will reduce the animal's energy the quickest and so do most damage because it moves the parameters fastest. Air speed is more important than moisture, which is more important than temperature.

Table 8 shows even a small increase in air speed at animal height increases the LCT significantly and is even more of a thermal stress than lying on a wet bed. Higher air speeds above 2 m/s such as in a draught will increase the negative effect (Fig. 36). Use windbreak material to protect openings and stop this (Fig. 37).

Fig. 36: A calf or heifer lying behind this door will be exposed to a draught, which will significantly increase the thermal stress to which the animal is exposed.

Calf weight (kg)	LCT (°C) at two different air speeds	
	0.2 m/s	2.0 m/s
35	+9	+17
50	0	+9
100	-14	-1

Table 8: The effect of air speed and calf weight on the LCT of a calf[8]

Temperature

Calves and heifers maintain a constant body temperature, by maintaining their internal temperature within strict limits regardless of the external environment. They will work to maintain their body temperature when the outside environment moves outside this range. This range of temperature is called the thermo-neutral zone (TNZ), which is defined as the range of temperature within which the animal uses no additional energy to maintain its body temperature. This range is bounded by the LCT and upper critical temperature (UCT). For a newborn calf the range is 15–25°C, for a calf over 3 weeks of age the range is 6–25°C, and for a growing heifer 0–23°C.

The effective ambient temperature (EAT) is the actual temperature felt by the animal, which may be very different from air temperature. For example, the EAT of a calf housed in a clean, dry hutch, bedded with straw may be 10°C warmer than the air temperature.

Fig. 37: Rubber sheeting has been attached to the bottom of this building door to stop a draught affecting the heifers behind it

Conversely, the EAT of a heifer exposed to wind and rain may be considerably lower than the ambient temperature.[10]

The lower critical temperature (LCT) of a calf or heifer is the temperature below which it has to divert additional energy towards keeping warm. At this point, less of the feed consumed is available for growth. This reduces daily live weight gain because feed efficiency is reduced. Prolonged exposure to ambient temperatures below LCT becomes a physiological stress so an animal is more likely to succumb to disease. Fig. 38 shows the effect of different temperatures on growth rate for a 10-day-old calf. Table 2, Chapter 3, page 41, shows the additional milk replacer or whole milk required to maintain growth rates at different temperatures.

Housing does not generally need to provide additional heating beyond the needs of the calf or heifer's own body mechanisms: it simply needs to maintain the thermo-neutral zone for them. This is done by managing moisture, fresh air, air speed, bedding and drainage as described previously. For this reason, and depending on how a house is managed, the temperature the animal feels is often different from the actual air temperature.

For calves, however, the requirement for additional heat in the calf house varies according to the calf/calves, weather and quality of calf house environment. The LCT of a calf up to 3 weeks of age is 15ºC, below this it needs to expend energy to keep warm. One way to help the calf is with a jacket (Fig. 40).

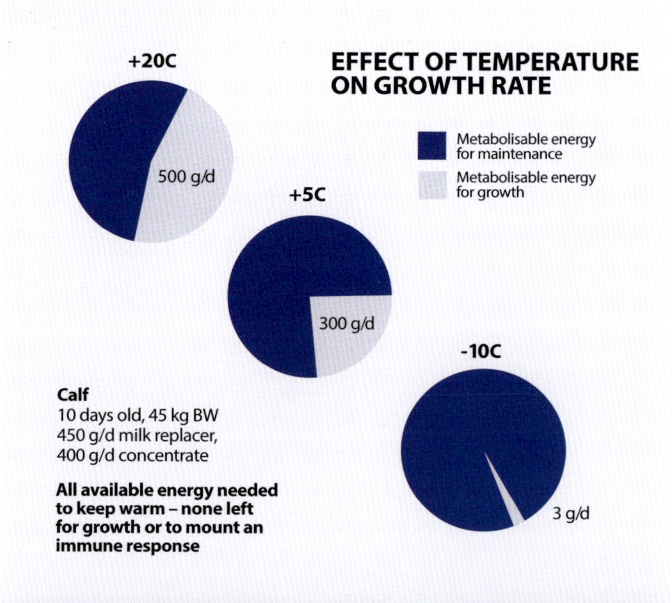

Fig. 38: Effect of temperature on growth rate.

If designed properly, the outer shell of the calf jacket will prevent outside moisture from penetrating the calf's hair coat – keeping it dry. The calf jacket insulation layer should be made of moisture-wicking material, channelling moisture away from the calf's body and preventing sweat from building up and getting the calf's coat wet (Fig. 39).

Fig. 39: Choosing the right calf jacket.

Comfort is very important – check the fit of the calf jacket. The front closure needs to stay securely closed, and straps should be tightened to fit the size of the animal without interfering with freedom of movement. Test for a proper fit by putting a hand between the strap and the calf's leg. Never use a calf jacket where the strap comes across the belly and could rub against the navel. This kind of irritation quickly leads to an infection.

Fig. 40: Calf jacket

Calf jacket protocol (Livestock Management Systems Ltd (2014)) :

- *Jackets must be made of breathable materials*
- *Locate a maximum/minimum thermometer in the calf house – next to the calves*
- *Reset the thermometer every morning during autumn and winter housing*
- *Decide on a set temperature for your system, e.g. 10°C*
- *Agree a starting protocol with staff e.g. three consecutive night-time minimum temperatures below 10°C = jackets go on*
- *Calves must be dried off before using a jacket*
- *Place a clean jacket on every calf below 1 week of age and on entry to the calf house*

Fig. 41: Calf heater

Calf jackets are equivalent to approximately one unit of "nesting score", i.e. nesting score 2 + jacket = nesting score 3 without jacket.[3] Removing jackets will depend on environmental conditions, but calves over 4 weeks of age do not need them.

- *Know the weather forecast – steady or rising temperatures is fine, but with falling temperatures beware of removing jackets too early*
- *Agree with staff a temperature protocol for healthy calves, e.g. 1 week old >10°C OK; two weeks >5°C OK; more than 3 weeks > 2°C OK*
- *Refer to recent night-time minimum temperatures*
- *Consider individual calf condition – feed intake, health, growth rate*
- *Remove jacket in the morning not the afternoon*
- *Remove dirt from jacket with hose/light power wash or pre-soak*
- *Wash jacket a 40-50°C with detergent and dry*
- *Using calf jackets is NOT an excuse to use less bedding!*

Calf heaters or heat lamps are another useful way to provide additional heat in the calf house. Most frequently used for sick calves, they can also be used to raise the temperature in the pens of healthy calves (Fig. 41). Heaters can be used to warm individual pens, or hung above group pens, where the younger or smaller calves will lie underneath them and older calves not needing the additional heat can lie elsewhere (Table 9). Plug heat lamps into a thermostat that allows them to come on at a certain temperature, e.g. in the calf house below 10°C.

Calf heater protocol
Provide a waterproof ring main and sockets (IP66 standard) above pens
Provide a frost stat on the circuit; ensure the power rating can accommodate all the heaters on the circuit
Use 250 w infra red bulbs over individual pens (one per two pens)
Use 1500 w linear heater (IP55 standard) wall mounted or hanging above group pens
Hang linear heater at height recommended by manufacturer
Agree a set temperature for frost stat; override for sick calves

Table 9: Calf heater protocol[11]

High air temperature and sun heat increases skin temperature enough to cause discomfort (EAT can increase by 3–5°C in sunlight). Black coats absorb more heat than mostly white coats. Rumination also increases body temperature.[4] Evaporation of body heat from a calf or heifer occurs via water loss from sweat and breath (panting). If temperatures lead to increases above the UCT, then appetite is reduced and water requirements increase. This too leads to reduced daily live weight gain and increased risk of disease.

When considering heat stress, not only is temperature important but also relative humidity. Calves and heifers can tolerate higher temperatures if relative humidity is lower. The temperature–humidity index (THI) is commonly used when evaluating heat stress in dairy cattle and is expressed as a single value representing the combined effects of both temperature and humidity. In humid climates, such as the UK, it is humidity that is the limiting heat stress factor.[12] Although used as a reference for heat stress in dairy cows, the temperature-humidity Index (Fig. 42) is still relevant for dairy calves and heifers.

	Relative Humidity %								
	20	30	40	50	60	70	80	90	100
22°C	66	66	67	68	69	69	70	71	72
24°C	68	69	70	70	71	72	73	74	75
26°C	70	71	72	73	74	75	77	78	79
28°C	72	73	74	76	77	78	80	81	82
30°C	74	75	77	78	80	81	83	84	86
32°C	76	77	79	81	83	84	86	88	90
34°C	78	80	82	84	85	87	89	91	93
36°C	80	82	84	86	88	90	93	95	97
38°C	82	84	86	89	91	93	96	98	100
40°C	84	86	89	91	94	96	99	101	104

No heat stress | Moderate heat stress | Severe heat stress | Dead cows

Fig. 42: Temperature-Humidity Index for dairy cattle (Adapted from Frank Wiersma, University of Arizona, 1990)

There are many inexpensive temperature and humidity readers available that will show current and maximum-minimum readings over time (Fig. 43). Place one in calf and heifer housing, near to the calves and record daily – during both winter and summer months.

Calves, in particular, may have a hard time coping with hot weather. With indoor housing, forced ventilation with fans, open curtain walls, and natural convection will move air (Fig. 44). Where calves are housed in hutches, ensure that ventilation apertures are open to create air movement and ventilation (Fig. 45). Use hutches that are opaque and made of UV-resistant material. The temperature in these types of hutches will not increase rapidly, giving the calf a chance to adapt. Hutches should always be equipped with adjustable top and back openings to facilitate good draught-free air movement to ensure proper ventilation. Raising the hutches off the ground to ensure ventilation at bedding level is another option. Provide clean fresh water at all times. If not, growth rates will decrease. It is also a good idea to take advantage of the cooler morning and evening temperatures when the calves are more energetic and have better appetite and accommodate feeding times accordingly.

Fig. 44: Forced ventilation with fans to reduce heat stress

Fig. 45: Ensure that ventilation apertures of hutches are open to create air movement and ventilation when it is hot

Fig. 43: A maximum–minimum thermometer located next to the calf pens

Light

There must be adequate lighting in a building to allow cattle to be inspected at any time. Cattle cannot be kept in permanent darkness. If natural light is insufficient to meet the needs of cattle inside, then artificial lighting must be provided. However, there must be an appropriate period of rest from artificial lighting[1]. Cattle are daytime creatures. For heifers up to calving either natural light – with 10% of the roof as translucent sheeting, or artificial light at 100–200 Lux at animal level to aid observation, is appropriate[4] (Figs 46 & 47).

Fig. 46: Placing a new roof on an old shed with translucent sheeting, plus opening up old windows in the walls, has provided adequate natural lighting for the calves in this shed

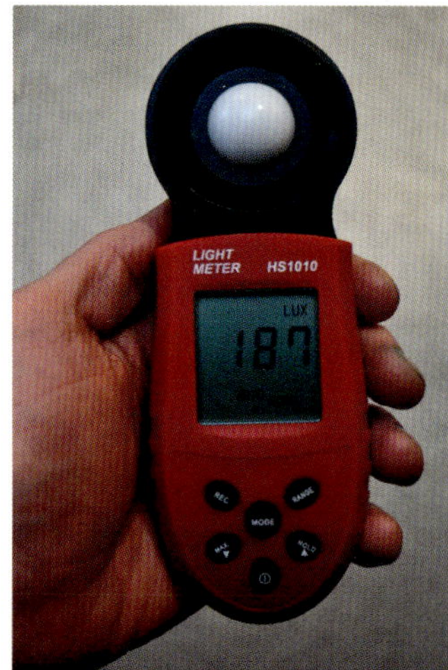

Fig. 47: A Light meter measuring light in a building in Lux

Plentiful light doesn't just make things easier for working with calves; it also has significant effects on the calves. As the hours of daylight decrease, the deposition of body fat increases. Insufficient light is interpreted by the calf as winter conditions. In response the animals start to lay down body fat. Increase hours of daylight to 16 using artificial light, as this promotes efficient use of nutrition. Less fat will be laid down in favour of building more muscle protein, so increasing growth rates.

Fig. 48: All heifers in this group have access to feed at the same time

Space requirements for feed and water

Newborn calves should have access to clean, fresh water from birth. Young calves prefer warm water. Starter grain should be available to nursing calves within 4 days of birth. Space requirements for feed and water for nursing calves fed from buckets in individual pens is obvious. It is more complicated with groups of calves and heifers.

Fresh feed delivery is the primary motivator for cattle to come and eat. With feeding of concentrates to young calves or precision feeding of older heifers, all members of a group should have space to eat at the same time to ensure that the smallest, most timid individuals have similar access to feed as larger or more high-ranking heifers in the group. Groups of heifers fed a total mixed ration must have adequate space for the entire group to eat all at once to prevent timid individuals from having access to feed only after it has been sorted by higher-ranking heifers[2] (Fig. 48).

The feeding area must be of an appropriate throat and table height for the size of calves or heifers using it (Table 10 & Fig. 49). Dividers or headlocks along the feed face will deter "boss" individuals from pushing lower-ranking heifers away from feed (Fig. 50). Slat bar feed fences allow the use of a taller neck rail to avoid potential injuries from heifers reaching for feed (Fig. 51). Post-and-rail feed barriers should only be used for heifers older than 8 months, and must be placed lower to prevent heifers from walking out of the pen[2] (Fig. 52). Animals should be able to pass behind those already feeding without disturbing them, which requires a passage at least 2 m wide.[1]

<table>
<tr><th>Body weight (kg)</th><th><60</th><th>60</th><th>100</th><th>150</th><th>200</th><th>300</th><th>400</th><th>500</th></tr>
<tr><td>Approximate age – Holstein (months)</td><td>1</td><td>1.5</td><td>3</td><td>5</td><td>7</td><td>11.5</td><td>15.5</td><td>20</td></tr>
<tr><td>Feeding space, width per head (cm)</td><td>30</td><td>30</td><td>36</td><td>41</td><td>51</td><td>56</td><td>61</td><td>66</td></tr>
<tr><td>Feed manger floor height above the litter alley floor (cm)</td><td colspan="8">10</td></tr>
<tr><td>Throat height above the litter alley floor (cm)</td><td>33</td><td>33</td><td>36</td><td>37</td><td>38</td><td>42</td><td>46</td><td>50</td></tr>
<tr><td>Neck rail height above litter alley floor (cm) with slant bar feed barrier</td><td colspan="2">76</td><td>86</td><td>97</td><td>104</td><td>112</td><td>122</td><td>127</td></tr>
<tr><td>Neck rail height above litter alley floor (cm) with post-and-rail feed barrier</td><td colspan="3">71</td><td>76</td><td>81</td><td>97</td><td>112</td><td>117</td></tr>
<tr><td>Water trough perimeter per head (cm)</td><td colspan="4">5</td><td>6</td><td>8</td><td>9</td><td>91</td></tr>
<tr><td>Maximum water height (cm)</td><td colspan="3">–</td><td colspan="2">74</td><td>79</td><td>84</td><td>86</td></tr>
</table>

Table 10: Feed and water space requirements for calves and heifers by weight[2]

Fig. 49: The throat and table height is far too high for these buffalo heifers, making it awkward for them to eat properly. Intakes and subsequent growth rates will be affected

Fig. 50: Heifers in headlocks at a feed fence prevents bullying of less dominant individuals

Fig. 51: Slat bar feed fence

Fig. 52: The rail is too low in this post-and-rail feed face, meaning these 4-month-old heifers are having to reach for feed and are rubbing their necks – like the third calf from the left. If it were raised higher the heifers might walk out. This type of feed fence should, therefore, only be used in heifers 8 months and older.

Adequate trough space or water bowls must be provided to allow at least 10% of a group to drink at anytime.[1] If calves or heifers are brought back to a pen from an extended time away from water, it is even more essential to ensure adequate space for many to drink at the same time, particularly in times of heat stress. Ensure water troughs are not too tall. Small calves may have difficulty reaching water if the trough is placed on the feed table.[2]

Do not locate water troughs next to bedding or at the back of pens where drinking animals can spill water and lead to it accumulating and causing damp bedding (Fig. 53). It is best to locate water troughs or bowls towards the front of a pen over concreted or free-draining areas (Fig. 54).

Table 10 shows feed and water space requirements for calves and heifers by weight. In practice it is often difficult to achieve specific requirements due to animals growing and/or a mixture of age groups. If this is the case, then allocations should be based on the average weight of the animals in the group, but monitored very closely to ensure all individuals have good access to feed and water.

Give consideration to where feed, water, feeding equipment and cleaning facilities are to be located. This does is not necessarily have to be in the same building, but there must be a provision for storage of feed, space for feed preparation and an area to allow cleaning of feed and water equipment with warm running water (Fig. 55).

Fig. 53: Water trough located above a bedded area causing bedding to become damp

Fig. 55: Storage of feed, feeding equipment and cleaning facilities

Fig. 54: Tipping water trough located at the front of the pen over concrete, so any spillage can drain away. It also means the water can be tipped onto the concrete to allow the trough to be cleaned

Cleanliness and hygiene

Cleaning is very important, requiring the right materials and the right amount of time to do the job. Poor hygiene for any cattle is unacceptable. The most important part of a good hygiene policy, especially in modern production systems, is time – not only to carry out the work but also to allow floors, pens, fittings and utensils to dry after wetting. A short dry period helps kill bacteria and viruses that survive the active cleaning process, and is a vital part of biosecurity[8] (Fig. 56).

Fig. 56 Cleaning calf facilities

Design for managing heifers

Each calf and heifer group should have access to a restraint system. Vaccination, reproduction, hoof trimming and other management schedules need to be considered and accounted for in housing. A crush or automatic head catch with a gate can alleviate the need to hold a calf or heifer into a small area while attempting to halter and tie her head (Fig. 57). While young heifers will certainly need observation and sometimes a little training to use headlocks along a feed face, this method of restraint is a more efficient use of time and labour, resulting in the restraint of a group of calves or heifers in a short amount of time, and produces much less stress and frustration to animals and people for vaccinations, breeding, reproductive examinations and other management activities (Fig. 50). These programmes are more likely to be accomplished in a timely and consistent manner, resulting in improved health and performance, if they can be done efficiently and conveniently.[2]

Fig. 57: A calf crush for tasks such as ear tagging, disbudding, castration and vaccination

References

1. Agriculture and horticulture development Board and Wm Morrison Supermarkets plc. (2012) Dairy housing – A Best Practice Guide.

2. School of Veterinary Medicine, University of Wisconsin-Madison. (2015) *The Dairyland Initiative*. Available at: https://thedairylandinitiative.vetmed.wisc.edu Accessed February 2015.

3. Nordlund, K. (2014) Preferred features of barns associated with improved calf health. *Cattle Practice* 22.

4. Charlton, S. J. (2009) *Calf Rearing Guide*. Context Publishing.

5. Davis, C. L. and Drackley, J. K. (1998) *The Development, Nutrition and Management of the Young Calf*. Ames (IA) Iowa State University Press.

6. The Danish Agricultural Advisory Centre. (2001) *Housing Design for Cattle – Danish Recommendations* (3rd Edition).

7. Livestock Management Systems Ltd. (2013) *Buildings Symptoms and Solutions*.

8. Agriculture and Horticulture Development Board. (2013) Better management of bovine respiratory disease (pneumonia). Better Returns Programme.

9. Lago, A., McGuirk, S., Bennett, T., Cook, N. and Nordlund, K. (2006) Calf respiratory disease and pen microenvironments in naturally ventilated calf barns in Winter. *Journal of Dairy Science 89*.

10. Calf Notes. Com. (2000) Calf Note #59 – Environmental Effects on Calf Feeding – Basic Concepts. http://www.calfnotes.com/pdffiles/CN059.pdf. Accessed March 2015.

11. Livestock Management Systems Ltd. (2014) Calf jacket/heater protocol

12. Pedersen, S. (2014) Effects of heat stress in cattle. *Veterinary Times 43*.

Getting the heifer in calf

Veterinary Guide To Rearing Dairy Heifers

6

Minimal losses, optimal growth rates and excellent fertility management are key factors of a good dairy heifer-rearing system.[1] Dairy heifers only start to produce a return on investment at first calving. Length of the non-productive rearing period is largely governed by farmer decisions on plane of nutrition and reproductive management.[2] Promoted "goals" for Holstein heifer fertility are:

- ***Reach puberty at 9–11 months***
- ***Start breeding at 13–14 months***
  - ***375–400 kg (55–60% of mature body weight)***
  - ***129cm withers height***
- ***Calve at 24 months***
  - ***580–600 kg (85% of mature body weight)***

Why calve at 24 months?

In practice, few farms achieve an average Age at First Calving (AFC) of 24 months. In the UK the average AFC is 26.2 months.[3] So why is it so important to calve at 24 months?

1. The optimal age at first calving (AFC) of 24 months minimises costs of rearing:

   – The main contributors to total costs are feed (35.6%), labour (24.7%) and bedding (8.9%)[2]

   – For each extra day in age at conception the average cost is £2.26.[2]

2. Heifers in block-calving herds must calve at 23–26 months or they will not be able to join the herd

3. Age at calving and herd culling rate both have a large effect on the number of replacements needed and the cost being carried[4] (Table 1):

   – UK dairy herds have an annual culling rate of 22–25%.[5] In a 100-cow herd with a culling rate of 23%, calving heifers at 24 months of age, required 12 fewer replacements than a herd with the same culling rate but calving at 30 months

AFC (months)	**Herd culling rate (%)** assuming 5% calf mortality						
	17	20	23	26	29	32	35
22	33	39	44	50	56	62	67
24	36	42	48	55	61	67	74
26	39	46	52	59	66	73	80
28	42	49	56	64	71	78	86
30	45	53	60	68	76	84	92

Table 1: The total number of heifers needed by a 100-cow herd at different culling rates and AFC[4]

4. The future performance of heifers that calve at 24 months is much better:

a. They have more days in milk and higher life-time milk yield

   – Cows calving at 23–25 months produce almost 7000 kg more milk over 5 years compared to those animals calving for the first time >30 months[3]

b. They have better fertility throughout their life

   – Lower subsequent calving intervals

   – Higher conception rates[3]

c. They are more likely to survive longer in the herd[3] (Table 2)

d. This future productivity, fertility and survival are all key determinants of profitability

While other aspects of heifer rearing are critical, few have the economic ramifications of inadequate growth and delayed AFC.

		23–25 months	>30 months
Lactation 1	305 d yield (kg)	8811	8914
	Calving interval (days)	402	451
Lactation 2	305 d yield (kg)	9908	9633
	Calving interval (days)	392	434
Lactation 3	% entering lactation 3	70	50
	Milk per day of life (kg)	11.8	9.0

Table. 2: Relationship between AFC and subsequent performance[3]

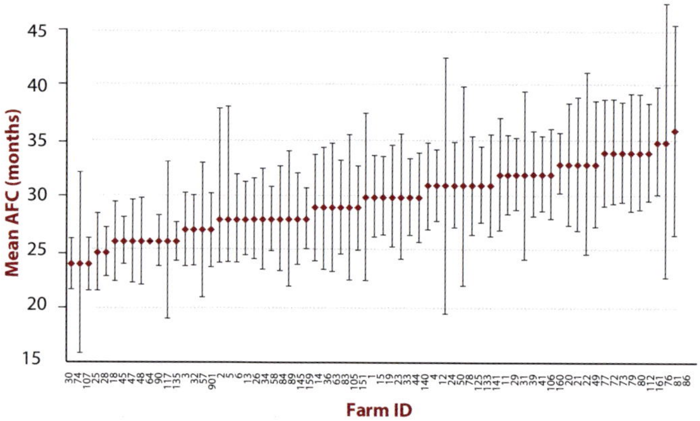

Fig. 1: Variation in average AFC and range of AFC on a selection of farms from the author's veterinary practice

While a focus on an average AFC is essential so is the range of AFC within a herd. Variability in growth rates (often due to suboptimal nutrition and/or disease), and fertility within groups of heifers, can lead to a large spread in age at which heifers calve.[6] In Fig. 1, some herds are achieving an average AFC close to 24 months, but have heifers calving at much younger and older ages than this. For example, herd 117 has an average AFC of 26 months, but a range from 19 to 33 months – this is not a successful programme.

In Fig. 2, although average AFC is close to ideal at 26 months, over half the heifers are calving at more than 26 months – around 54%. So distribution of AFC must be taken into account as it is more representative of the real reproductive performance in heifers than simply looking at the average AFC.

The major factors driving AFC are:

- *Age at puberty*
- *Growth patterns*
- *Management factors*
  - *Voluntary waiting period*
  - *Pen management*
  - *Pregnancy checks*
- *Oestrus detection efficiency*
- *Breeding management*
  - *Synchronisation*
  - *Fixed time artificial insemination*
  - *Natural service*
- *Conception rate*

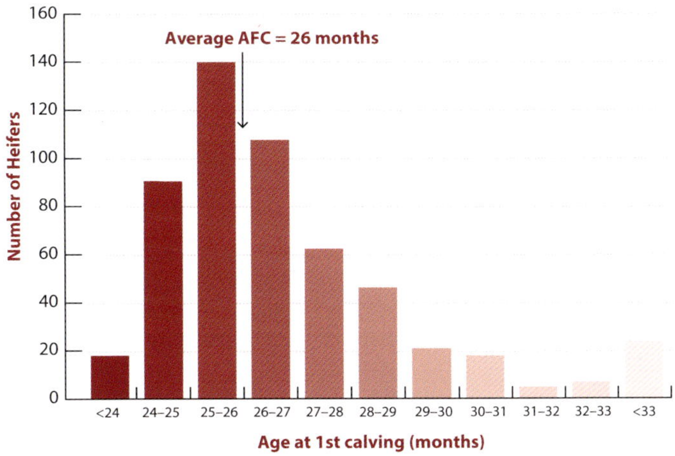

Fig. 2: Distribution of AFC among heifers in a herd

Fig. 3: The "Forgotten heifer" – often located far away in distant pastures or pens

The importance of good heifer reproductive management

The demands of a modern dairy farm often mean not enough time is spent when it comes to heifer reproductive management. Common reasons for this include:

- *Heifers are generally located far away in distant pastures or pens – the "forgotten heifer" (Fig. 3)*
- *Obsolete structures or cumbersome adaptations used for heifer housing*
- *Personnel with little or poor training to handle heifers*
- *Little or no breeding and performance records for heifers.*[7]

Early heifer growth and development along with efficient reproductive management are critical:

- *Without proper early development, breeding can't begin early*
- *Without good heifer reproduction the benefits of good rearing cannot be captured*

Gestation length is fixed so AFC relies on when first breeding starts, combined with the reproductive efficiency of the animal. The decision on when to start breeding is a management one, based on the age of the heifer, which is influenced by growth during the rearing period[3].

Heifers achieve puberty at more than 40% of their mature weight, and should reach 55–60% of their mature weight before first service. Body weight should have increased to 85% by first calving. To calve at 580–600 kg at 24 months of age, it is necessary to start first breeding at 13–14 months at about 375 kg. An average daily gain (ADG) of at least 0.8 kg/day throughout the rearing period is necessary to meet these targets.

In order to determine if the heifer is heavy enough weigh scales (Fig. 4) can be used. Alternatively, at this age withers height is strongly correlated to weight – 129 cm at 375 kg weight. Therefore, a line can be drawn on the wall at 129 cm so that when the heifer passes it can be determined if she is ready to serve (Fig. 5). Other methods for measuring withers height include a mark on the crush or a measuring stick.

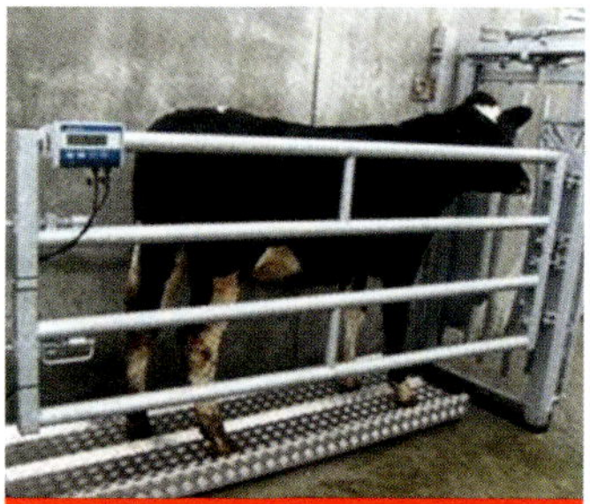

Fig. 4: Weighing a heifer on an electronic weigh scales. (www.agriband.ie)

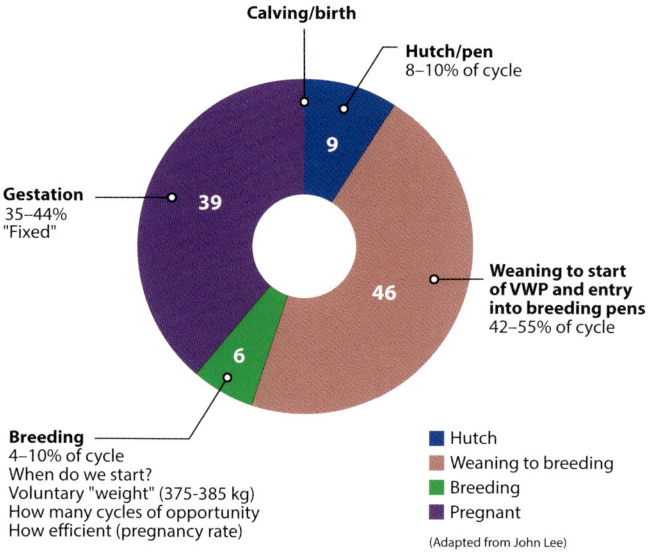

Fig. 6: Life cycle (birth to calving) of a heifer

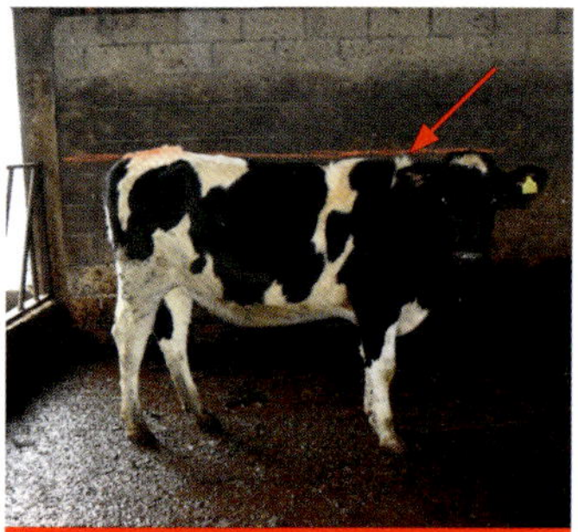

Fig. 5: The mark on the wall is at 129 cm – when heifers walk past their withers height can be determined, which gives an indication of whether they are ready to serve

Fig. 6 shows the life cycle of the heifer from birth to calving. Gestation length is fixed, and birth to weaning is a relatively short period of time and fairly consistent. The biggest times that can be influenced to achieve target AFC are weaning to breeding, and the breeding period itself.

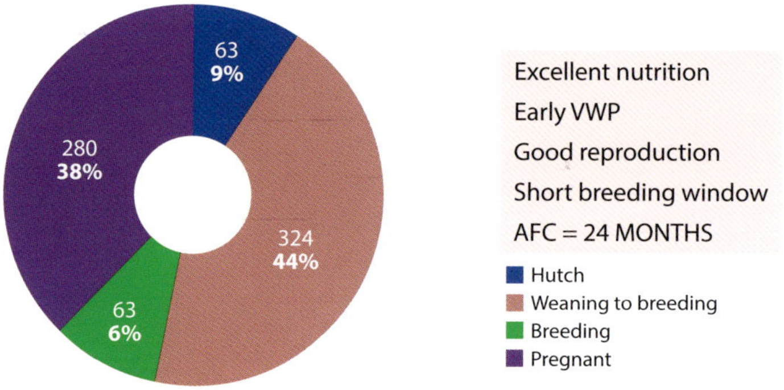

Fig. 7: Comparison of good heifer reproduction management, achieving an AFC of 24 months – "Program one"

Figs 7 and 8 are a comparison of good heifer reproduction management, achieving an AFC of 24 months – "Program one" and average heifer reproduction management, achieving an AFC of 29 months – "Program two".

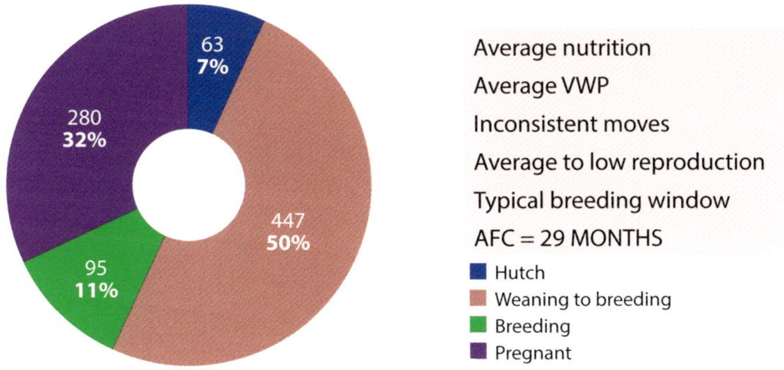

Fig. 8: Average heifer reproduction management, achieving an AFC of 29 months - "Program two"

To achieve AFC of 24 months as in Program one, a focus on heifer fertility management is essential to avoid time, productivity and profits being lost.

Strategies and management practices at breeding time pre-breeding

1. Heifer growth should be monitored at key points – birth, weaning, 6 months, 12 months and bulling. (Table 16, Chapter 3, page 77).
2. Development of the uterus should be checked before the heifer enters the breeding pen. "Reproductive tract scoring" should be performed by a vet on all heifers before breeding starts (Table. 3). Manage heifers with poor or immature uterine tract scores separately (cull or improve plane of nutrition) in order to improve uterine score and allow them to enter the breeding pen. Once above score 3 heifers can be moved to the breeding pen.[7]
3. Body condition should be monitored. Target body condition score (BCS) is 2.5–3.0 throughout the rearing process. Lower BCS is likely to indicate suboptimal nutrition or disease. Heifers that are fat (higher BCS) may lay down fat in the mammary gland, with reduced future milk production. Monitor and record BCS of all heifers every time they are routinely handled. Make any necessary adjustments, such as separating individuals from the group to increase plane of nutrition, as soon as a problem is identified.
4. Infectious disease represents a particular hazard to heifers bred for the first time and often managed away from the adult herd (therefore, not exposed to pathogen so that they build up natural immunity). Bovine viral diarrhoea (BVD) may cause poor conception rates, early embryonic death or abortion. Similarly, all other commonly found UK abortion causing organisms may affect heifers. Although less common, the bull may introduce infections such as *Campylobacter*.[9]

   Effective biosecurity measures will minimise the risk of introducing new infectious diseases, particularly *Salmonella Dublin* and *Campylobacter*. A herd health plan should be drawn up with the vet to protect the health of the breeding heifers according to known risks in the herd. This might include identification of cows carrying *Neospora* so that they are not used to breed heifers; identification and removal of BVD persistently infected cows (PIs); and completion of the appropriate vaccination programmes, e.g. against BVD, *Leptospirosis*, IBR and *Salmonella*, at least 1 month before service.[9]

   Not an infectious disease but "Tyre wire disease" or "Hardware disease" can be fatal to a heifer and yet the simple administration of a magnet at the same time as vaccination dramatically reduces the risk
5. Trace-element status of heifers should be checked. Low vitamin E / selenium status or low copper/high molybdenum may be associated with reduced fertility. These deficiencies may not show up in the adult herd that are receiving supplementary feed.
6. Move heifers to a breeding pen 2 weeks before first AI target. Move heifers into the breeding pen weekly, once they have achieved appropriate height = 129 cm and weight = 375–400 kg. This relies on:

- *Proper feeding and health management*
- *Good recording and compliance to management lists for moves.*

RTS	Uterine horn	Ovary			
		Length (mm)	Height (mm)	Width (mm)	Ovarian structures
1	Immature<20 mm diameter, no tone	15	10	8	No palpable structures
2	20–25 mm diameter, no tone	18	12	10	8 mm follicles
3	25–30mm diameter, slight tone	22	15	10	8–10 mm follicles
4	30 mm diameter, good tone	30	16	12	>10 mm follicles, corpus luteum possible
5	>30 mm diameter, good tone, erect	>32	20	15	>10 mm follicles, corpus luteum present

Table 3: Reproductive tract score system[8]

Oestrous detection

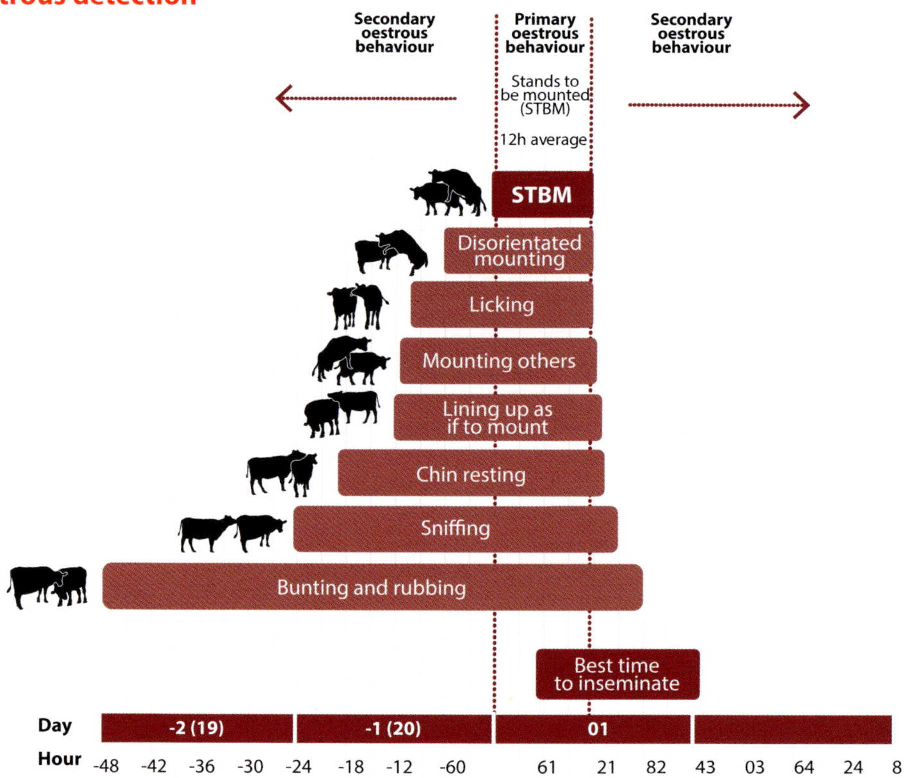

Fig. 9: Heifers showing signs of oestrus. (Adapted from Borsberry[10])

Oestrus detection is observing heifers for signs of oestrus or "bulling" (Fig. 9). This then allows heifers to be served at the correct time. Begin oestrus detection 3 weeks before the start of the service period, then any heifers not seen bulling can be examined to ensure their ovaries are active[9] (Table 3).

Time spent on oestrus detection in heifers is often limited, so a structured approach must be taken. The best times to observe oestrus are.[11]

- *At dawn and dusk when heifers are at grass*
- *At night and mid-afternoon when heifers are housed*
- *When heifers are stirred up, e.g. moving pastures*
- *After feeding when non-oestrus heifers lie down and those on heat are able to find each other, and so become more obvious.*

The most important factors determining oestrus activity levels are:[11]

- *The number of sexually active heifers in the group – activity peaking with three to five in oestrus*
- *The space and opportunity for sexually active animals to interact*
- *The ambient temperature – very cold and very hot conditions reduce oestrus behaviour*
- *Floor and foot conditions – slippery floors reduce oestrus behaviour*
- *Day length and light intensity – greater levels of activity are recorded where there are more hours of daylight and where the period of light has been extended artificially*
- *Nutrition – heifers with poor BCS may fail to show signs of oestrus*

Heat detection rates of only 40% are achievable in the course of routine work. In order to achieve higher rates, frequent, dedicated observations throughout the day during the breeding season are required[11] (Table 4).

NUMBER OF OBSERVATIONS	HEAT DETECTION EFFICIENCY (%)
1 or 2	49
3	54
4	61

Fig. Table 4: The impact of the number of observations on heat detection[11]

Guidelines for better heat detection are:[11]

1. *Set aside time specifically for heifer observation and heat detection separate from time spent feeding, scraping, etc.*
2. *Combine quiet management activities, such as cleaning dung from cubicle kerbs, liming cubicles, moving electric fences and grassland recording with heat detection, always creating reasons to get among heifers.*
3. *Undertake three 30 min.+ observation periods spaced as evenly as possible through the day – 6am, 2pm and 10pm – with particular emphasis on the night-time round for heat detection rate targets of 70% or more.*

Although heifers usually show strong signs of bulling, lasting 12–18 hours, they may not be observed because they are housed or at pasture away from the adult herd.[9] In these circumstances, the use of oestrus detection aids can be used to help in oestrus detection but should never be used as a substitute for good husbandry.

An ideal oestrus detection aid should be:[11]

- *Able to provide continuous heifer surveillance*
- *Accurate – combining low levels of missed heats and false positives*
- *Simple – so it gives real savings in management time*
- *Good value for money*
- *Accepted by staff as an aid rather than a threat*
- *Practical for the conditions under which it is being used.*

Fig. 10: Large freeze brand to allow easy identification of individual heifers

Oestrus detection aids:

1. *Records, diaries and charts*
   - *The most fundamental aid to heat detection, allowing heifers likely to be in heat at a particular time to be observed closely*
   - *Calendars and diaries (paper or electronic) allow key fertility events and details to be recorded*
   - *Action lists – of animals due for service and possible return dates*
   - *3-week breeding diaries – allowing likely next heat and service dates to be predicted from the recording of observed heat activity and providing particular help in reducing false positives*
2. Large ear tags, freeze brands and collars (Fig. 10)
   - Essential to have a good highly visible means of identifying animals and vital in minimising false positives
3. Vasectomised bulls
   - Equipped with a halter-mounted, chin-ball marking device that leaves paint marks on the rump of heifers in heat (Figs 11 & 12)
   - They tend to mount non-pregnant animals only
   - Their presence improves detection by increasing vigour of heats and heifers on heat mingle actively with the bull
   - Vasectomised bulls can be used from 350 kg live weight until 2½ years of age when they tend to be too big for the cubicles
   - 80–100 heifers are an effective working limit for one bull and it is important to top up the marker every week
   - Problems can arise with excessive weight gain, loss of libido, inherent dangers of a bull (particularly one frustrated by continual removal of heifers going for AI) and the potential spread of venereal disease[11]
4. Penned bulls (Fig. 13)
   - Bulls penned in a suitable location, which heifers pass closely and regularly, are valuable in stimulating oestrus behaviour
5. Tail paint and heat mount detectors
   - Tail paint, Kamars and Estrotects (scratch cards) are all useful aids to determine oestrus activity (Figs 14–16).
   - Each is applied to the tail head area
   - Tail paint is removed following mounting, while Kamars and Estrotects are stimulated to produce a bright colour
6. Motion sensing systems
   - Oestrus is associated with increased activity, so activity meters can also be used to identify bulling heifers, once activity exceeds a certain threshold
   - 3 distinct concepts are marketed:

   i. Pedometers – strapped to the leg of an animal, which measures overall animal movement and detects increases during oestrus (Fig. 17)

Fig. 11: Chin ball marking device for a vasectomised bull

Fig. 12: Chin ball marking device for a vasectomised bull

Fig. 13: Penned bull

Fig. 14: Tail paint

ii. Necometers – attached to the neck of an animal, which detects by comparing a previously measured standard level of individual heifer movement to the increased movement exhibited in oestrus (Fig. 18)

iii. Optical sensors – compare captured images of individual animals with a reference image held on a computer database[11]

- Heifers don't pass through a parlour twice a day where traditionally activity can be downloaded into a sensor. So options include having the sensor placed over the water trough, or a radio-transmitter from which signals can be downloaded.[9]

Herds utilising consistent routines of oestrus detection such as visual observation or with the help of oestrus detection aids tend to have excellent oestrus detection (inseminated / eligible to AI) results of 65–70% or more.[7]

Conception rates

- *To calve at 24 months means beginning to serve at 13 months*[12]
- *The target is to have at least 60% of heifers served by 15 months of age*[12]
- *First-service conception rate among Holstein heifers is 67% in the United Kingdom*[13]
- *The major factors that will influence conception rate are:*
  -Oestrus detection accuracy
  -AI technician
  -Sire
  -Animal, including breed and age
  -Days since last insemination at pregnancy diagnosis
  -Semen – conventional vs sexed
  -Season, i.e. heat stress

It can be common to hear from farmers: "I get 90% of my heifers pregnant in 3 services" (Table 5):

SERVICE	CONCEPTION RATE %	PERCENTAGE PREGNANT (CUMULATIVE)
1	60	60
2	55	82
3	50	91

Table 5: Conception rates as reported on farm

The results in Table 5 look impressive but the problem is that no time value is included. What percentage of heifers are inseminated every 21 days? Hand-in-hand with efficient heat detection goes efficient insemination of heifers to achieve high and early conception rates. Insemination rate is the area of greatest opportunity in heifer reproductive management.

20% of UK farms exclusively use natural service with a bull, 29% do not use a bull at all, with the remaining 51% starting with AI then switching to a bull later.[2] Table 6 compares the advantages and disadvantages of service by a bull or artificial insemination.

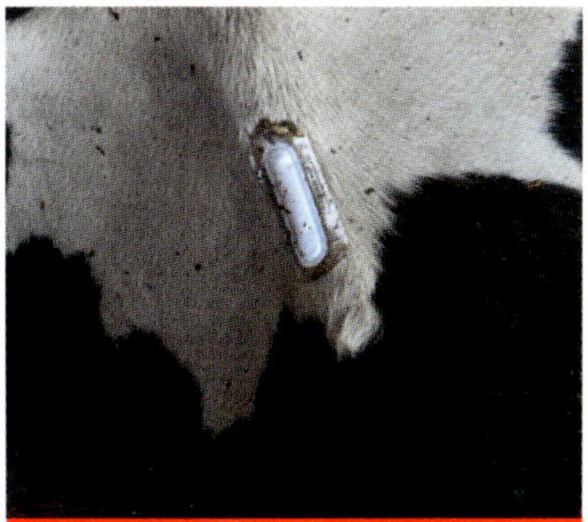

Fig. 15: Kamar

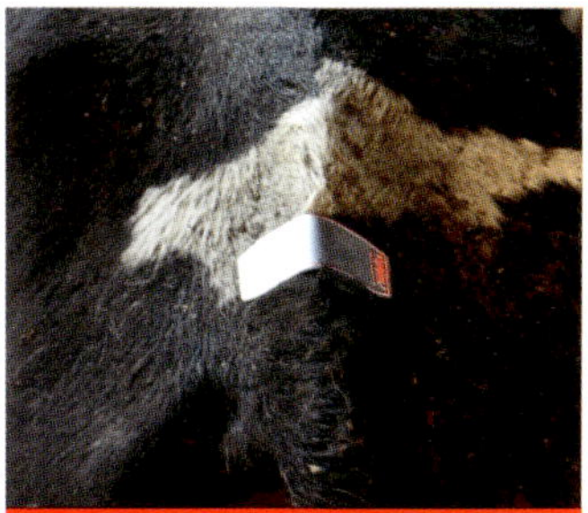

Fig. 16: Estrotect

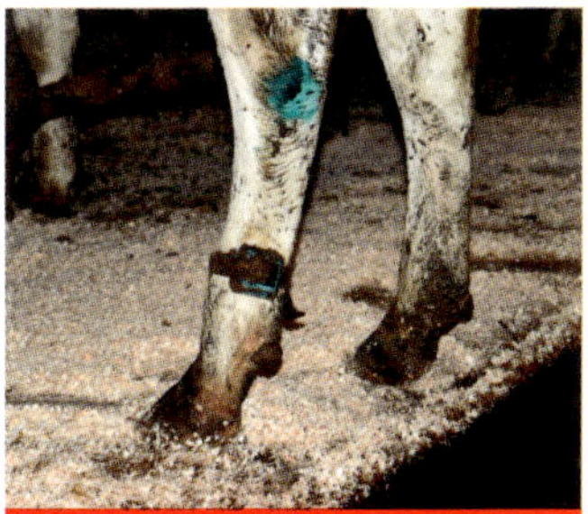

Fig. 17: Pedometer

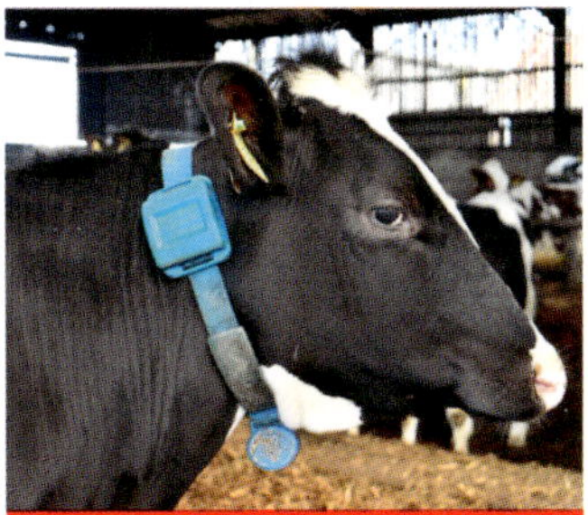

Fig. 18: Necometer

	ADVANTAGE	DISADVANTAGE
AI	Allows greater control over genetic selection including calving ease	Requires more labour and management input and infrastructure
	Will allow faster genetic improvement of the herd	Higher cost*
	Avoids problems of inbreeding	
	Easier to control, manipulate and monitor fertility	
Bull	Low labour input	Genetic potential often not known
	Optimum heat detection efficiency and multiple serves per heat should produce better pregnancy rates than AI	More risk of infectious disease, particularly a hire bull
		Bull prone to risk of problems with fertility
		Risk of trauma to heifers
		Health and safety aspects of working with a bull

*In smaller herds a bull may be more expensive to run

Table 6: Advantages and disadvantages of service by the bull or artificial insemination (Modified from Bazeley and Hayton[14])

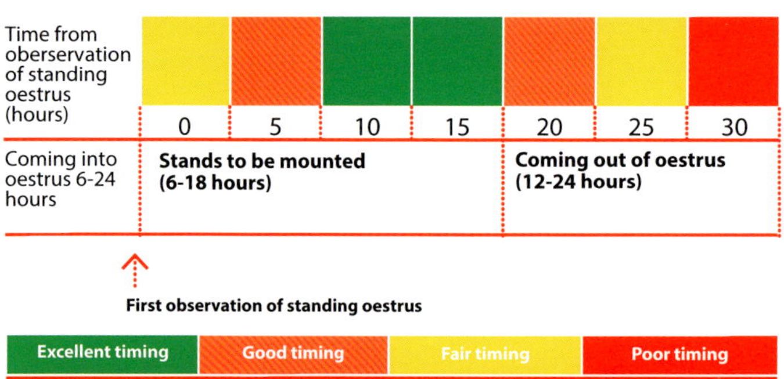

Fig. 19: Insemination timing guidelines[11]

The whole process of oestrus and AI is new to the heifer and so minimising stress during this time is crucial, otherwise she is unlikely to become pregnant. Reduce stress by:[11]

- *Maintain a familiar routine*
- *Separate the heifer from the group for as short a time as possible*
- *Provide good holding facilities with access to feed and water.*

The advantages of AI far outweigh those of a bull – for a progressive modern dairy farm aiming to continually improve the genetic status of the herd, AI is the best option. Semen should be from tested bulls to foster or maintain genetic progress. Avoid relying on a bull to breed heifers.

Best conception rates are achieved when heifers are inseminated at their most receptive time. Ovulation occurs about 12 hours after the end of the standing heat. Insemination very early in oestrus or after ovulation gives poor results. When a heifer is first seen in heat, however, there is no reliable way of knowing how long ago the oestrus began or how long it will continue.[11]

Timing of insemination should be based on observation of standing heat (Fig. 19). On this basis:

- *Heifers should be served within 24 hours of being first seen in oestrus*
- *When oestrus is first seen in the afternoon or evening, inseminate the next morning*
- *When oestrus is first seen in the morning, inseminate in the afternoon*
- *When an AI technician is used, all heifers seen in oestrus by the time the inseminator arrives should be inseminated*

If the heifer has to be held away from the group to await insemination for several hours, ensure:[11]

- *The rest of the group is not visible*
- *There is adequate, familiar food and water*
- *Adequate space available to avoid bullying*
- *Individuals can be easily and calmly restrained*
- *Return to existing groups is done promptly.*

Good inseminators place the semen just through the cervix into the uterus with a margin of error of less than 1 cm. Consistently achieving this accuracy requires:[11]

- *Restraining the heifer from sideways movement*
- *Heifer and inseminator at the same level*
- *No restrictions to inseminator movement*
- *Insemination facilities undercover, with unobstructed heifer entry and exit routes.*

The correct handling of semen is vital to minimise damage that can affect conception rates. Most importantly:[11]

- **Removing semen**
  - *Keep time spent removing semen from the liquid nitrogen to a minimum*
  - *Temperatures can reach 10ºC in the tank neck and damage can be caused if the canister of semen is raised above the frost line (Fig. 20)*
  - *If the correct semen cannot be located within 10 seconds, lower the canister into the flask again for a further 10–15 seconds*
- **Locating semen**
  - *The location of the semen in the tank should be noted before removing the plug*
  - *A tank inventory aids location, plus a means of stock control, with straws marked off after each withdrawal*
- **Thawing semen**
  - *Thaw semen in a water bath at 35-37ºC for at least 40 seconds (Fig. 21). Calibrate thermometers regularly to maintain accuracy*
  - *Dry semen straws completely before use*
  - *Remove straws with tweezers (not fingers) and flick to remove any trapped nitrogen*
  - *15-20 straws can be thawed at the same time for multiple inseminations. More water may be required plus agitation*
  - *Number of straws thawed should not exceed number that can be used within 10–15 minutes*
  - *To avoid cold-shocking the semen keep the insemination gun warm using a paper towel and close to the body (under clothes)*
  - *Cut the straw at the crimped end and place a plastic sheath over the straw and gun to keep it clean and dry (Fig. 22)*
  - *Inseminate the semen as soon as possible after thawing (Fig. 23).*

Following AI heifers should be pregnancy tested early: 30–35 days (Fig. 24). This will identify any problems with conception or early embryonic death before empty heifers have lost months when they are assumed pregnant.

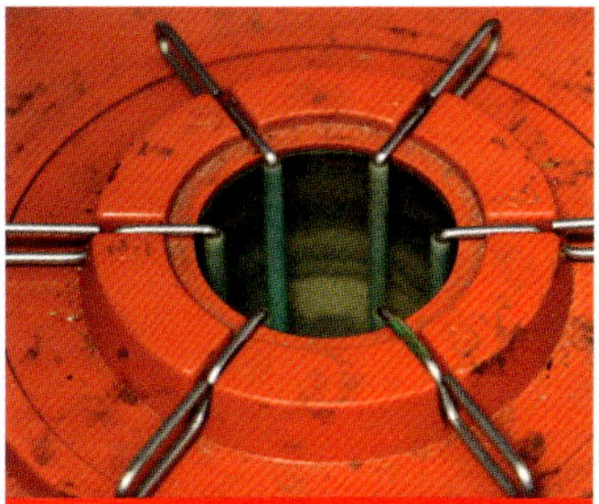

Fig. 20: The frost line in the neck of a liquid nitrogen flask used for storing semen

Fig. 21: Thawing semen straws in a water bath

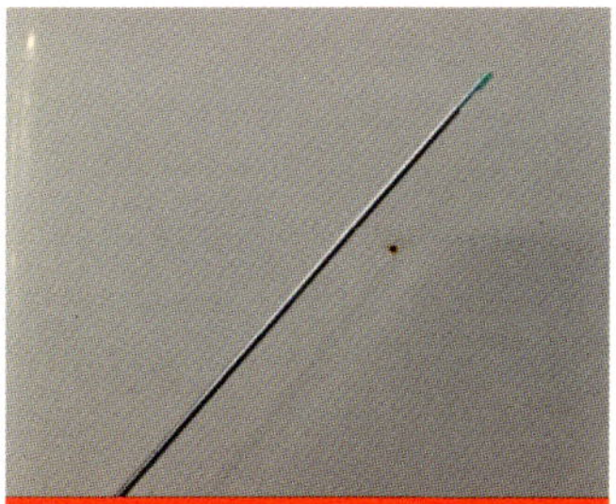

Fig. 22: Place a plastic sheath over the straw and gun to keep it clean and dry

Fig. 23: Inseminating a heifer

Fig. 24: Pregnancy testing a heifer

Sexed semen

Calves are valuable carriers of genetic progress. Besides helping to improve the genetic potential of the herd, heifers also give birth to other calves that are more productive and live longer than the calves born from cows that have had several calves. When pregnancy coincides with lactation, the competition between the mammary gland and the placenta leads to an offspring with a lower milking ability and lower longevity.[15] Heifers out of heifers are the most productive and long-lived animals in the herd.

Sexed semen produces an average of 90% heifer calves. Other advantages include:[11]

- *Concentrating dairy replacement breeding on fewer, better cows*
- *Producing fewer low-value purebred dairy bull calves*
- *Producing more higher value beef cross calves*
- *Reducing calving difficulties by breeding dairy heifers from heifers*

The average first service conception rates are lower for sexed semen at 59.2%[2] vs 67%[13] for conventional semen. This could be due to a reduction in the number of sperm resulting from the sexing process, which is insignificant when the insemination process is handled well but leaves fewer margins for error in semen handling, thawing and insemination.[11]

Manipulating oestrus

Intending to serve earlier does result in earlier first calving – but a farm's actual first service age is consistently 2 months higher than intended.[12]

Synchronisation can improve heifer breeding results by reducing AFC and the requirement for oestrus observation.[9] It can allow seasonal calving herds to calve as many animals as possible at the beginning of the calving season (front loading), and enables a batch of heifers to calve together so that they can be managed as a group that join the milking herd together.[16]

18% of farms in the UK use synchronisation to help serve heifers. Those that do, serve on average 1.5 months sooner.[12]

Choice of approach depends on:

- *Number of heifers to be dealt with*
- *Handling facilities for AI and subsequent calving*
- *Labour availability for handling and heat detection*

Success relies on:

- *Well-grown cycling heifers*
  - *Any that are not observed in oestrus should be checked to make sure their ovaries are active, they are not pregnant or are not freemartins*
- *Good handling facilities*
  - *Sufficient to handle the insemination of all heifers within 2–4 hours*
- *Good nutritional management before and afterwards*
- *Proper semen handling and AI technique* (see Conception rates (Page 133))
- *Good records and control of treatments*
- *Careful planning and agreement between everyone involved well ahead of time*

When implementing an oestrus synchronisation protocol it must be followed precisely. Each product must be administered at the correct dose on the correct day (refer to a protocol sheet) and in some cases at the right time of day. In addition, oestrus synchronisation products must be stored, handled and administered correctly.[17]

Should a mistake occur in product administration or the treatment timeline, seek advice from your vet. To minimise the probability of making a mistake, write each of the days of treatment, the product name, dose to be administered and day of insemination on a calendar and ask your vet to review it before beginning the protocol.[17]

A number of synchronisation programmes are available:

1. *Prostaglandin (PGF2a) alone (Fig. 25)*

   • *All heifers in the group are injected with prostaglandin PGF2a, then inseminated to observed oestrus*

   • *All heifers that have not been inseminated 11 days later are injected again with PGF2a and again inseminated to observed oestrus*

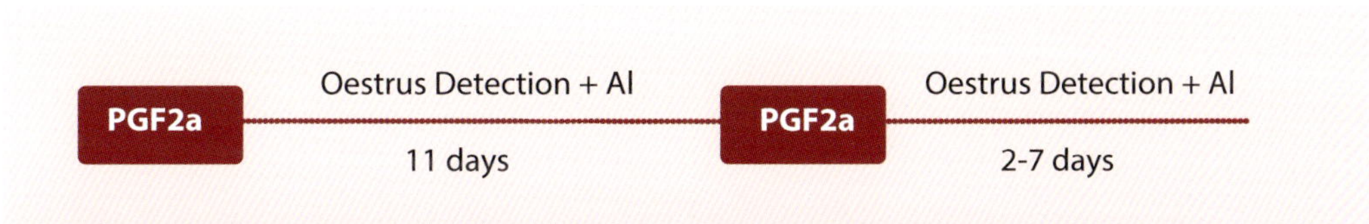

Fig. 25: Prostaglandin synchronisation

Use of PGF2α in cycling heifers at the beginning of the breeding period will increase the proportion of heifers inseminated within a week, compared to non-treated heifers. It will also induce more synchronous oestrus behaviour, aiding heat detection, because more animals are in oestrus simultaneously. It reduces time to pregnancy by 7–10 days in dairy heifers compared to untreated controls.[18]

A period of intense observation is required if heifers are to be served to observed oestrus – a minimum of one person per 100 heifers; so fixed time AI programmes may be preferred.[9]

2. *Intra-vaginal progesterone-releasing (P4) device plus Prostaglandin (Fig. 26)*

   • *Day 0 P4 device in*

   • *Day 6 PGF2a*

   • *Day 7 P4 out*

   • *56 hours later fixed time AI (FTAI)*

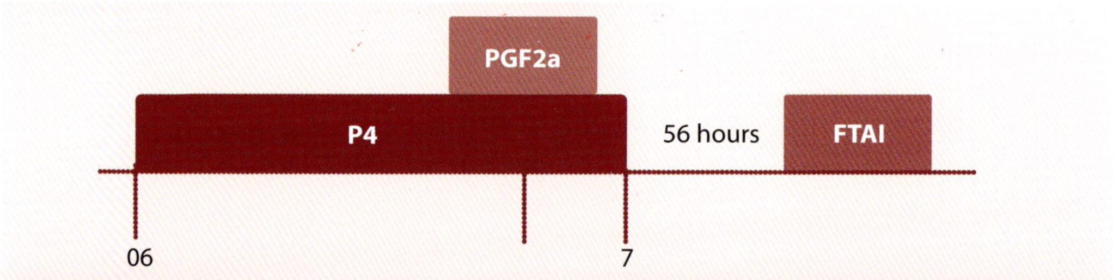

Fig. 26: Progesterone + Prostaglandin synchronisation

Heifers synchronised with an intra-vaginal progesterone-releasing device plus PGF2α become pregnant at a faster rate than those synchronised with PGF2α alone.[19] It also results in a tighter synchrony of oestrus than PGF2α alone.[18]

3. *5 day Progesterone program (Fig. 27)*

- *Day 0 P4 device in*
- *Day 5 P4 out + PGF2α*
- *Day 8 GnRH (Gonadotrophin Releasing Hormone) + FTAI*

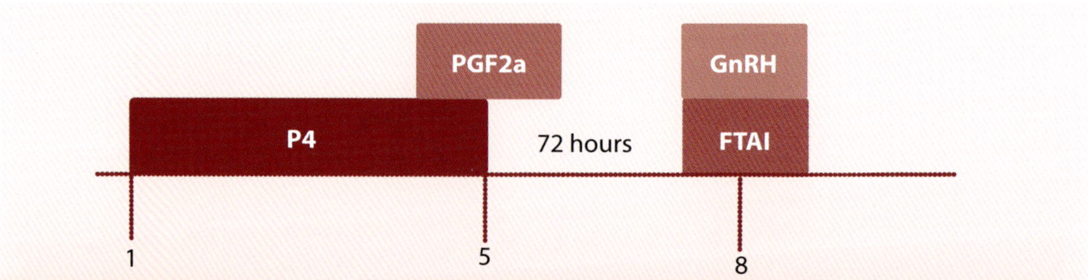

Fig. 27: 5 day Progesterone program

The use of GnRH stimulates ovulation. This more simple option of a 5-day programme is easier in terms of management and less costly. Good conception results can be expected to timed AI breeding.

4. *Modified 5 day progesterone program (Fig. 28)*

- *Day 0 P4 in + GnRH*
- *Day 5a.m. P4 out + PGF2a*
- *Day 5p.m. PGF2a*
- *Day 8 GnRH + FTAI*

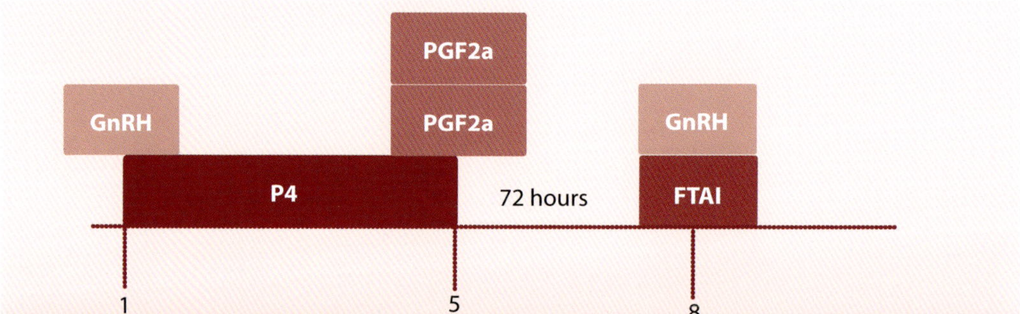

Fig. 28: Modified 5 day Progesterone program

A second PGF2α injection is required to ensure complete luteolysis of the newly formed corpus luteum produced after the GnRH treatment given at the P4 device insertion. Good conception results similar to conception achieved by AI following natural oestrus are attained.

Which synchronisation programme do you choose?

There are many more synchronisation programmes for heifers beyond the 4 most common mentioned above. The decision should be taken in conjunction with your vet and be the programme that best fits your breeding programme, facilities and personnel. Potential approaches are:

- *Use PGF2α on day of entry into the breeding pen*
- *Second dose 11 days later if not bred*
- *At 21–28 days post-entry, check those that are not yet inseminated*
  - *Remove freemartins/abnormal tracts*
  - *Enrol on modified 5-day progesterone programme*

Alternatively:

- *Use modified 5-day progesterone programme for all heifers up front*
- *Aggressive oestrus detection*
- *Prompt pregnancy diagnosis at 30–35 days post-AI*
- *Use of PGF2α to all open heifers*

There will be another period of intense effort as synchronised heifers calve, with extra requirements for housing and handling facilities, and to introduce them to the milking routine[9]. These requirements must also be addressed before embarking on a synchronisation program.

When to stop trying to get the heifer in calf

When a considerable amount of time and effort has gone into rearing a heifer and then trying to get her in calf, it is very difficult to draw the line and take the decision to either sell or cull her when she fails to get in calf. Have clear guidelines of when to stop attempting to breed heifers:

- *Cull any heifer that aborts*
  - *Heifers that abort are 2.7 times more likely not to finish their first lactation, compared to those that do not abort[20]*
- *Cull after 4 unsuccessful services or no more than 6 cycles of breeding opportunity*
  - *Heifers needing more than 4 inseminations to conceive are half as likely to finish the first lactation compared with heifers that become pregnant at first service[20]*
- *Don't turn open heifers into a bull pen after AI*
  - *Often resulting in keeping heifers that would have otherwise been culled.*

References

1. Brickell, J. S., McGowan, M. M. and Wathes, D. C. (2008) Aspects of heifer rearing. *Cattle Practice* 16.
2. Boulton, A. C., Rushton, J. and Wathes, D. C. (2015) The management and associated costs of rearing heifers on UK dairy farms from weaning to conception. *Open Journal of Animal Sciences* 5.
3. Cooke, J. S. and Wathes, D. C. (2014) Rearing heifer calves for optimum lifelong production. *Cattle Practice* 22.
4. DairyCo Desktop Review. (2005) Heifer rearing in the 21st century.
5. Esslemont, R. J. and Kossaibati, M. A. (1997) Culling in 50 dairy herds in England. *The Veterinary Record* 140.
6. Ettema, J. S. and Santos, J. E. (2004) Impact of age at calving on lactation, reproduction, health and income in first-parity Holsteins on commercial farms. *Journal of Dairy Science* 87.
7. CEVA Animal Health: reprodAction Scientific Meetings. (2013) Reproductive management in dairy heifers.
8. Andersen, K., LeFever, D., Brinks, J. and Odde, K. (1991) The use of reproductive tract scoring in beef heifers. *Agricultural Practice* 12.
9. Bazeley, K. And Hayes, E. (2014) Fertility management of dairy heifers. *Livestock* 19.
10. Borsberry, S. (2001) *Bovine Reproduction, A Practical Guide*. Intervet UK Ltd.
11. DairyCo (2012) Managing heat detection. Section 6 PD+ Fertility Improvement Programme.
12. Atkinson, O. (2015) Dairy Youngstock Project – Wales. Dairy Development Centre.
13. Brickell, J. S., Bourne, N., McGowan, M. M. and Wathes, D. C. (2009) Effect of growth and development during the rearing period on the subsequent fertility of nulliparous Holstein–Friesian heifers. *Theriogenology* 72.
14. Bazeley, K. J. and Hayton, A. J. (2007) *Practical Cattle Farming*. Crowood Press.
15. Gonzalez-Recio, O., Ugarte, E. and Bach, A. (2013) Trans-generational effect of maternal lactation during pregnancy: A holstein cow model. PLoS ONE 7.
16. Breen, J., Down, P., Kerby, M. and Bradley, A. Restoring the dairy herd: rearing youngstock and replacing cows in *Dairy Herd Health*. Green, M. (2012) CABI.
17. Smith, M., Perry, G., Atkins, J., Jinks, E., Pohler, K. and Patterson, D. (2012) Keys to a successful oestrus synchronisation and artificial insemination program. *American Red Angus Magazine* 4.
18. Stevenson, J. L., Rodrigues, J. A., Braga, F. A., Bitente, S., Dalton, J. C., Santos, J. E. P. and Chebel, R. C. (2008) Effect of breeding protocols and reproductive tract score on reproductive performance of dairy heifers and economic outcome of breeding programs. *Journal of Dairy Science* 91.
19. Lucy, M. C., Billings, H. J., Butler, W. R., Ehnis, L. R., Fields, M. J., Kesler, D. J., Kinder, J. E., Mattos, R. C., Short, R. E., Thatcher, W. W., Wettemann, R. P., Yelich, J. V. and Hafs, H. D. (2001) Efficacy of an intravaginal progesterone insert and an injection of PGF2α for synchronising oestrus and shortening the interval to pregnancy in postpartum beef cows, peripubertal beef heifers, and dairy heifers. *Journal of Animal Science* 79.
20. Bach. A. (2011) Associations between several aspects of heifer development and dairy cow survivability to second lactation. *Journal of Dairy Science* 94.

Dealing with disease

Veterinary Guide To Rearing Dairy Heifers

7

Dealing with Disease

Disease is an unfortunate reality when rearing heifers. Disease costs time and money, it reduces growth rates and, consequently, the age at when a heifer enters the adult herd. Disease can lead to others in a group suffering, and in some circumstances death of one or many heifers. Disease is totally demoralising and depressing problem to have to deal with.

However, because disease is an almost inevitable fact of rearing heifers you must be prepared. It is important to know how to recognise disease, what has caused it, how to treat it and how to prevent it happening in future. It is also very important to know what "acceptable" levels of disease are, and when to seek expert veterinary help.

Sadly, there are many different diseases that can affect calves and heifers. This chapter will focus on the main diseases. It is worth remembering if you hear hooves behind you - it is more likely to be a horse than a zebra, i.e. common things occur commonly! But, if in doubt always seek veterinary help.

Calf scour

Every farmer will have encountered an outbreak of scour (Fig. 1). It is the most common cause of disease and death in pre-weaned calves,[1] accounting for up to 50% of all causes of death.[2] While mainly a problem pre-weaning, scour can also occur post-weaning.

In severe cases an outbreak of scour can kill up to a third of affected calves. But calf deaths only represent a small proportion of the financial damage an outbreak can cause. Scour will typically cost at least £44 per calf affected, excluding labour.[3] In addition, there is a delay in reaching age at first calving and lower yields in first lactation.

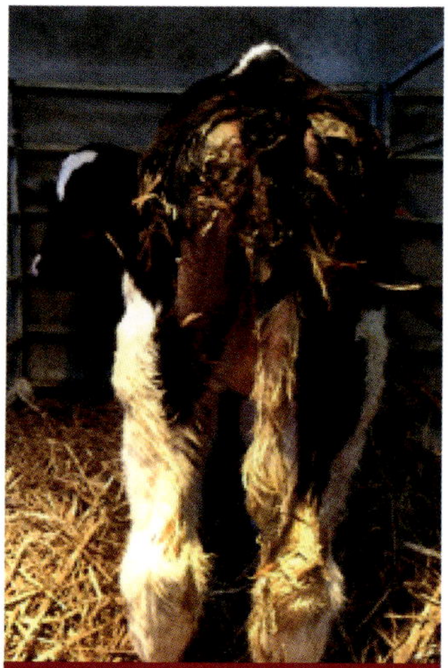

Fig. 1: A scouring calf

TARGETS

- 24 hours – 60 days: <25%
- 61–120 days: <15%
- 121–180 days: <1%

Causes

There are many infectious and non-infectious causes of scour, and as clinical signs are general, it is not possible to identify or predict a specific cause from clinical signs alone. The infectious bacteria, viruses and protozoa that cause scour are important, but it must be remembered they can be found on farms with and without clinical disease. Furthermore, mixed infections with one or more agents are common. Outbreaks are multifactorial and can be caused by an interaction between the environment, management, feeding and the micro-organisms.[4] An understanding is therefore needed of why an outbreak has occurred, as much as the infectious agents involved.

Although many infectious agents have the potential to cause scour in calves, at the moment 6 micro-organisms are significant:

- Viruses: Rotavirus and Coronavirus
- Bacteria: *Enterotoxogenic E. coli* (ETEC) and *Salmonella* (particularly types *Dublin* and *Typhimurium*)
- Protozoan parasites: *Cryptosporidium* and *Coccidia*

Fig. 2 shows their frequency of diagnosis in the United Kingdom between 2006 and 2013.

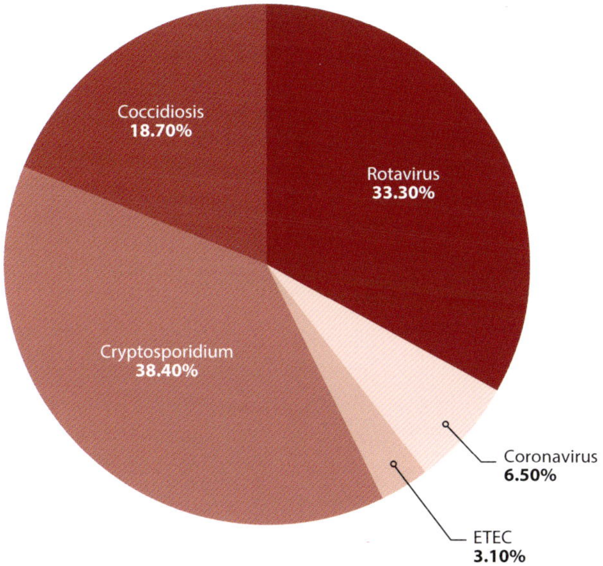

Fig. 2: Scour diagnoses in calves in Great Britain between 2006 and 2013, where a diagnosis has been reached[5]

Rotavirus and coronavirus

The most common source of infection is faecal contamination from cows, or other infected calves with scour. These viruses target cells lining the small intestine of the calf. They multiply within these cells and destroy them, so the calf is unable to digest the sugar (lactose) in milk, and also unable to fully absorb water, salts or nutrients. Diarrhoea is usually from 4 days to 3 weeks of age, with a peak incidence of 6 days for Rotavirus, and 7–10 days for Coronavirus. Older calves can sometimes be affected too. The ability of the intestine to replace these cells is not compromised though, so gradual replacement with new cells takes place over several days and the calf will recover after 5–7 days.[6]

Cryptosporidium

This protozoa causes scour in many species including man. The source of infection is adult cows. Damage caused is similar to that of rotavirus with disease occurring from 4 days to 3 weeks of age. The infective dose of cryptosporidium is very low, just five oocysts can result in infection. This low dose contrasts greatly with the parasite's ability to produce huge numbers of infective oocysts. Over the 7–10 day duration of a typical infection, infected individuals may shed 10,000,000,000 oocysts, with faecal oocysts counts in the order of a million per gram.[7] The oocysts are resistant to most commercially available disinfectants and can stay alive on the farm for many months, so hygiene is very important for control.[6]

Enterotoxogenic *E. coli* (ETEC)

A calf becomes infected with many different strains of *E. coli* within hours of birth. This constantly changing population of organisms inhabits the calf's intestines for life, which is normal and healthy. However, some strains such as *E. coli* K99 also known as enterotoxogenic *E.coli* (ETEC) behave differently. ETEC sticks to the lining of the intestine and produces toxins that reduce normal absorption of salts and fluids and stimulates secretion of body fluids into the gut that are then lost in scour.[6] The high levels of secretion lead to very watery scour and rapid dehydration. This severe scour occurs in the first 4 days of life and is often fatal.

Salmonella

Many *Salmonella* species can cause scour but *Salmonella Dublin* is the most common in the UK. Any age of animal can be affected, but disease usually occurs between 2 and 6 weeks of age. Scour is often profuse, watery and fatal (Fig. 3). *Salmonella Dublin* can also enter the blood stream causing systemic infections such as pneumonia, septicaemia and joint ill.

Fig. 3: A calf that has collapsed and is dying due to scour caused by *Salmonella Dublin*

Coccidia species

This protozoa causes scour between 3 weeks and 6 months of age. Infection can produce chronic scour, or severe bloody scour with mucus and straining to pass faeces. Only 5% of infected animals show clinical signs of coccidiosis, however,[8] while the remaining 95% are subclinical. It is subclinical coccidiosis that has greatest effects on a herd, with ill thrift and poor growth rates. Calves usually recover from scour within 10 days, but some severe cases have a prolonged recovery.

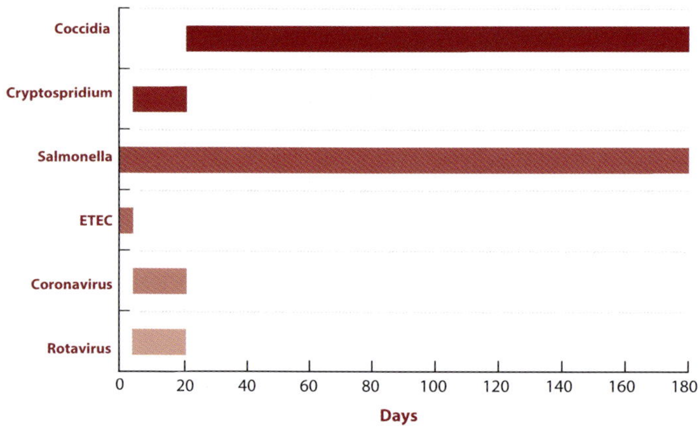

Fig. 4: Timing of occurrence of infectious scour organisms

Non-infectious scour is usually referred to as nutritional scour and can occur anytime pre-weaning (See Table 6, Chapter 3, page 53).

Effects on the animal

Scour is best considered a failure of intestinal uptake of salts and water sufficient to overwhelm the compensatory capacity of the colon.[9] In the healthy calf there is a flux of fluid and salts with net secretion occurring proximally and net reabsorption occurring distally, with 90% of net reabsorption occurring in the lower small intestine with the remainder being absorbed in the colon (large intestine). The colon operates at only about 40% of its absorptive capacity in the healthy calf and so has considerable functional reserve capacity. Scour is only seen when this reserve has been overwhelmed such that liquid faeces are produced. Consequently, by the time a calf has scour, the disease process has been in progress for a while.[9]

Risk factors

- *Unhygienic calving areas*
- *Poor colostrum management*
- *Inadequate feed*
- *Poor hygiene at feeding and of feed equipment*
- *Inconsistencies in feeding, e.g. timing, mixing, concentrations, volumes, temperatures*
- *Weaning*
- *Inadequate housing and poor housing hygiene (see Chapter 5)*
- *Climatic stress*
- *Mineral deficiencies*
- *Other diseases*
- *Poisons, e.g. antibiotics, arsenic, copper, fluorine, mercury, molybdenum, mycotoxins, sodium chloride and some plants*

Clinical signs

1. *Faeces / Dung*
   - *Increase in frequency and volume of faeces with high water content*
   - *Infectious scour – loose to watery, bright yellow or green, strong odour, can be blood-stained and/or contain mucous* (Fig. 5)
   - *Nutritional scour – pasty, soft, clay-like dung, yellow or white in colour*
2. *Posture*
   - *Dull/depressed – indifferent to other calves*
   - *Head down*
   - *Sometimes swollen or tucked up abdomen*
3. *Eyes*
   - *Dull, sunken* (Fig. 6)
4. *Coat condition*
   - *Dry, dull, stary coat* (Fig. 7)
5. *Temperature*
   - *Low <37.5ºC if very dehydrated*
   - *High >39.5ºC if bacterial infection*
6. *Appetite*
   - *Slower, reduced or none*

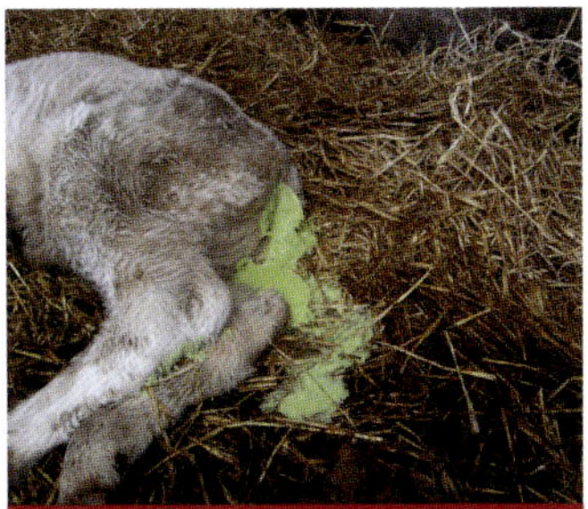

Fig. 5: Infectious scour. (Photo courtesy of C. Penny)

Fig. 6: Dull, sunken eyes

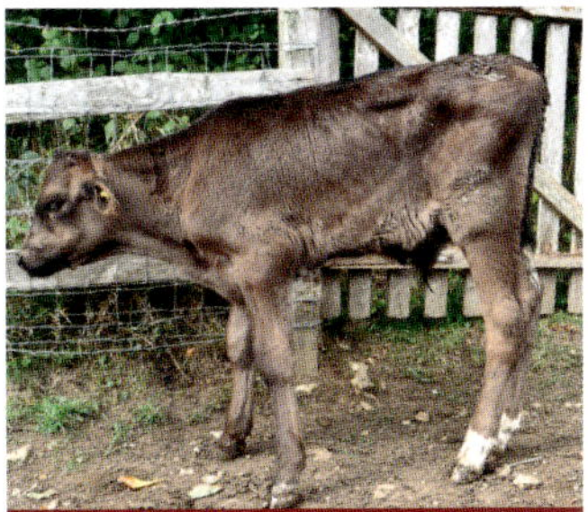

Fig. 7: Dry, dull, stary coat

Diagnosis

Diagnostic testing must be done in consultation with a vet, and will influence treatment, prevention and control strategies, which vary for the different infectious causes of scour.

Collection of six faecal samples (minimum 15 g each) from untreated calves in the early stages of disease, gives the best possible chance of a diagnosis. Samples must come direct from the rectum and not the floor where they would be contaminated. Samples should be tested for all potential causes of infectious scour – while taking into account age of animals affected and clinical signs.

If any calves have died then a post-mortem examination (Fig. 8) is an excellent way to reach a diagnosis – preferably at an approved laboratory where any necessary follow-up testing can be performed. Calves must have died less than 24 hours earlier, to give the greatest chance of reaching a diagnosis.

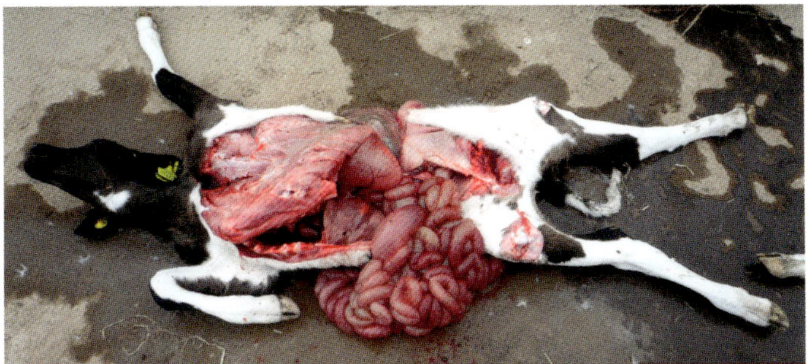

Fig. 8: An on-farm post-mortem showing the small intestine of this calf was inflamed and gassy due to scour

Treatment

The aim of treatment is to help the calf survive the episode of scour in as good a shape as possible. Treatment must focus on replacement of pathological losses, correcting acidosis, meeting nutritional requirements, TLC and, in some instances, the administration of antibiotics.[10]

Calves with scour lose an average of about 10% (but up to 20%) of their body weight in fluid in addition to normal losses.[11] For this reason oral rehydration therapy (ORT) is the single most important treatment (Fig. 9), and is usually successful if instigated immediately after the scour has developed.[1] ORT is formulated to correct or prevent water and salt deficits and metabolic acidosis. Oral rehydration solutions should contain:[11]

- *90–130 mmol/litre sodium*
- *Glucose and/or another facilitator for the absorption of sodium*
- *An alkalinising capacity of 60–80 mmol/litre from bicarbonate or bicarbonate precursors*
- *10–30 mmol/litre potassium*

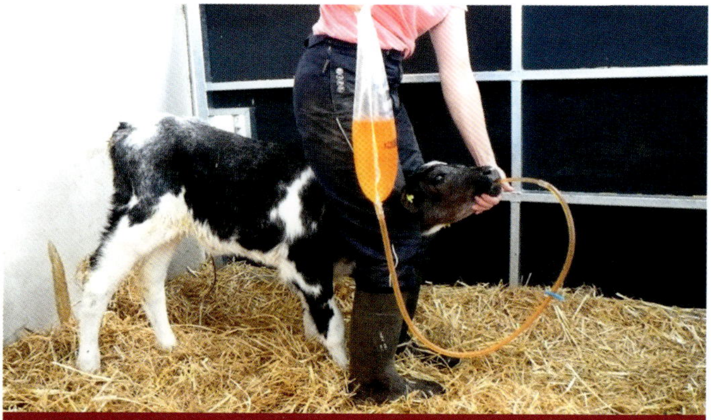

Fig. 9: Oral rehydration therapy in a scouring calf

Always continue to feed milk to the scouring calf. Milk has natural antimicrobial properties and is highly digestible, so is the ideal nutrient support to the scouring calf.[12] Continued milk feeding not only provides the energy required for weight gain and growth throughout scouring but also provides the nutrients that are necessary for the recovery of the lining of the intestine.[11]

Give two extra feeds of 2 litres ORT daily, with milk feeds in between. For example, in a twice-a-day fed calf: ORT – milk – ORT – milk. Each ORT feed needs to be separated from milk feeding by at least 2–3 hours; this is because ORT interferes with clotting of milk in the abomasum. This can be done for 3 days but, if no improvement is seen, seek veterinary guidance. Acetate-based solutions are the best choice for calves suffering with diarrhoea that are still receiving milk, because acetate does not interfere with abomasal milk clotting.[12]

A general rule is that if a calf is standing it should receive ORT and if it is collapsed (Fig. 3) or cannot stand without assistance then it should receive intravenous fluids by a vet. Exceptions are:

- *If the calf is dehydrating rapidly "eyes sinking as you watch"*
- *Severely acidotic calves with abomasal dilation as oral fluids remain in the atonic stomach without being absorbed*
- *Calves that are worsening despite ORT.*

It is often necessary to administer intravenous fluids (IF) on farm, but preferably patients should be hospitalised if facilities and biosecurity issues allow.

The use of antibiotics in calf scour is controversial. The majority of infectious agents involved are either viral or protozoal, and so do not need antibiotics. However, in calves with scour and severe systemic involvement, intercurrent disease is common and the risk of bacteraemia or septicaemia is increased.[10]

Feeding of antibiotics to calves as a method of diarrhoea prevention is not recommended. However, calves with diarrhoea and systemic signs of illness should receive injectable antibiotics targeted towards coliform bacteria in the blood (due to likelihood of bacteraemia) and the small intestine (due to bacterial overgrowth). A calf is bacteraemic when exhibiting inappetance, dehydration, lethargy or fever.[13] If in doubt discuss with a vet.

Calf scour causes abdominal discomfort – so give pain relief. A non-steroidal anti-inflammatory drug will decrease inflammation in the gut and reduce the effects of endotoxaemia and septicaemia. This should encourage feeding and aid recovery, but must only be administered when a dehydrated calf is receiving ORT. If the calf is hypothermic then the use of a heat lamp or a calf blanket will help recovery.

Some causes on infectious scour do have specific treatments licensed for them, for example, Halofuginone to treat Cryptosporidium. Halofuginone can be used as both a treatment and preventatively when prescribed by a vet. It reduces disease severity and oocyst output – minimising further spread. On a group basis, prophylactic use of Halofuginone can prevent further scour due to Cryptosporidium after a diagnosis has been made. The drug is given daily for 7 days. It has a narrow safety margin and is toxic at twice the therapeutic dose, so ensure the recommended dose is given. Calves that have had scour for longer than 24 hours should not be treated with Halofuginone, and dehydrated calves should not be treated unless on oral fluids.

If *Salmonella* or ETEC are identified then antibiotics should be prescribed by a vet. A diagnostic laboratory will perform antibiotic sensitivity testing in such cases to enable appropriate antibiotic selection.

If a diagnosis of coccidiosis is made specific anticoccidial drugs should be used under veterinary prescription. If creep feeding calves, then medicated feed is likely to be the most efficacious and simplest method of treatment.[6]

Control

Irrespective of the causal agent, the route of infection for infectious calf scour is the same – ingestion or inhalation of the organism from an environment that is heavily contaminated by faeces. Hygiene and cleanliness are paramount and along with good colostrum are the key to prevention and control of calf scour.

Good hygiene must start from the moment the calf is born as many infections are picked up from the calving area. Hygiene and cleanliness must be maintained throughout the calf-rearing period, with specific care being taken to clean and disinfect feeding equipment between feeds (Fig. 10) and to thoroughly clean the environment on regular occasions and between batches of calves (Fig. 11). Scour is 1.9 times as likely when calves are reared on wet bedding, and 0.6 times as likely when there is disinfection between groups.[4]

General farm disinfectants are usually satisfactory. However, Cryptosporidium oocysts are highly resistant and persist in the environment for long periods of time. Ammonia-based disinfectants will kill them, but produce highly irritant fumes, so can only be used when the building has been de-stocked. Alternatively KenoCox at a 3% solution or Sorgene 5 at a 4% solution is effective.[6]

Fig. 10: Keep feeding equipment clean

Vaccination is available for many infectious scour organisms specifically a vaccine for Rotavirus, Coronavirus and ETEC, and a vaccine is also available for *Salmonella*. Both vaccines require vaccination of the cow during the dry period or the heifer as she approaches calving (it is possible to also vaccinate healthy calves 3 weeks and older for *Salmonella*). A vaccine is not a substitute for good management and both vaccines rely on good colostrum management to ensure protective antibodies are passed from cow to calf. (See Pneumonia, page 154, for appropriate management of vaccines.)

Nutritional scour can be prevented by having a clear consistent protocol for feed preparation. All feed and equipment must be clean; milk must be mixed at the same concentrations and temperature and fed in the same volumes at the same times each day. Any changes must be gradual and separated by one week.

Fig. 11: Clean housing and the environment between batches

Pneumonia

Calf Pneumonia is a lifelong disease. Outwardly calves may recover, but the internal damage to the lungs is irreversible (Fig. 12) and will affect lifelong health and productivity.

21% of all the disease in pre-weaned calves is due to pneumonia. 22.5% of all calf deaths pre-weaning are attributed to pneumonia (Fig. 13), and 46.5% of deaths post-weaning.[14] Calf pneumonia has major impacts on the economic performance of cattle operations. This is due to direct costs of disease, death and treatment as well as long-term effects on performance.

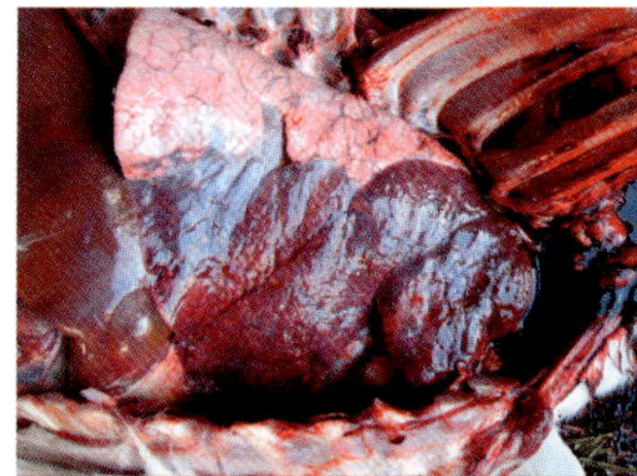

Fig. 12: The internal damage to a calf's lungs affected by pneumonia is permanent. The healthy lung in this image is the pale pink area at the top. The dark purple area below is compromised/damaged lung tissue caused by pneumonia

Fig. 13: Two calves that died as a result of a Mycoplasma pneumonia outbreak

- *A single case of pneumonia carries an average cost of £43.26 per sick dairy calf, and an average cost of £29.58 per calf for the rest of the group*[15]
- *Add the cost of a 2-week delay to first service,*[16] *at £1.65/day*[17]
- *Plus a 2.2% reduction in first lactation milk yield*[16] *= 165 litres (first lactation of 7503 litres*[18]*)*
- *With a current average UK milk price of 22.93 ppl (AHDB Dairy, February 2016 price), then the cost of a single case of pneumonia rises to £104.19*

Calves treated for pneumonia before 3 months of age are 2.5 times more likely to die after 3 months.[19] They are 2.4 times more likely to die between 3 months and 2½ years than heifers that had not been treated.[16] And they are 2.4 times more likely to experience dystocia and will calve 3 months later.[20] These estimates do not include a value for the welfare costs endured by the calves, nor the emotional cost to the stockman and others within the industry.[21]

Causes

Table 1 lists the main pathogens involved in calf pneumonia. While viruses and mycoplasma can be viewed as primary pathogens and bacteria secondary invaders, disease does not always follow that pattern.

TARGETS

- 24 hours – 60 days: <10%
- 61–120 days: <15%
- 121–180 days: <2%
- 6–12 months: <1%
- 12 months–calving: <0.5%

VIRUSES	
	Respiratory syncitial virus (RSV)
	Parainfluenza virus (PI3)
	Bovine herpes virus 1 (BHV-1)
	(Infectious Bovine Rhinotracheitis (IBR))
	Bovine viral diarrhoea
	Bovine adenoviruses
	Bovine coronaviruses
	Bovine rhinoviruses
	Influenza A virus
BACTERIA	
	Pasteurella multocida
	Manheimia haemolytica
	Histophilus somni
	Biberstinia trehalosi
	Arcanobacterium pyogenes
	Fusobacterium necrophorum
	Salmonella species (particularly Salmonella dublin)
	Mycobacterium bovis (TB)
MYCOPLASAMA	
	Mycoplasma bovis
	Mycoplasma dispar
	Ureaplasma species
PARASITES	
	Dictyocaulus viviparus
NON-INFECTIOUS CAUSES	
	Fog Fever
	Farmer's lung/diffuse fibrosing alveolitis
	Aspiration pneumonia

Table 1. List of the main pathogens involved in calf pneumonia

MANAGEMENT GROUP	CLINICAL SYNDROME	LIKELY PATHOGENS
Housed calves less than 6 months old	Severe acute pneumonia	RSV, PI3, *M. bovis, H. somni, M. haemolytica*
	Chronic coughing	Mycoplasma species
Housed weaned calves after summer	Severe acute pneumonia	RSV, PI3, IBR, *M. bovis, D. viviparus*

Table 2. Common clinical syndromes and casual organisms

Table 2 shows the common clinical syndromes and causal organisms in each case. It is important to make a diagnosis as this will affect potential treatment and vaccination options. As with scour though, pneumonia relies on an animal whose defences are weakened, an environment that is inadequate thus encouraging pathogen growth and subsequent high numbers to exploit the situation.

Risk Factors

- *Unhygienic calving areas*
- *Poor colostrum management*
- *Inadequate feed*
- *Poor hygiene at feeding and of feed equipment*
- *Inconsistencies in feeding, e.g. timing, mixing, concentrations, volumes, temperatures*
- *Weaning*
- *Inadequate housing and poor housing hygiene (see Chapter 5)*
- *Housing/assembling groups – housing from pasture; assembling in open courts; congregating around troughs or for shelter outdoors*
- *High levels of ammonia and/or dust in housing*
- *Climatic stress*
- *Mineral deficiencies*
- *Other diseases, e.g. coccidiosis*
- *Mixing/changing groups – calves from different sources; different ages; vaccinated and non-vaccinated; sick and healthy*
- *Moving – farm to farm or within the same farm complex; to and from shows*
- *Other stresses e.g. disbudding*

Effects on the animal

Different viruses and bacteria can affect the calf's respiratory system in different ways. The classic pattern is one of the viruses unlocking the door and then the secondary bacterial invaders rushing in and causing all the damage. However, viruses alone, or bacteria alone, can cause pneumonia, and often viruses can be found to "work together". Mycoplasma can cause pneumonia alone, in conjunction with viruses, or can act as a secondary invader.

The main processes that affect the respiratory system in pneumonia are:

1. *Ventilation of the air sacs in the lungs (alveoli) is impaired by respiratory depression; air and/or fluid accumulating in the chest around the lungs; accumulation of inflammatory cells within the air sacs; thickening and scarring of connective tissues of the lungs; and obstruction of higher airways in the lungs (bronchioles and bronchi)*
2. *Blood flow to the lungs is impaired by clots; thickening and scarring of connective tissues of the lungs; and accumulation of inflammatory cells in and around the blood vessels*
3. *Diffusion of gases across the air sac walls into the blood stream is impaired by fluid accumulation and air in the connective tissues around the air sacs (where it shouldn't be) causing enlargement and damage*

Diagnosis

The information gained from diagnostic investigations may have a limited impact on the management of a calf pneumonia outbreak as it happens, because a time delay will frequently mean the required information is not available at the time a treatment decision is made. However, providing a specific diagnosis of the cause of calf pneumonia outbreaks helps with future health-planning, disease prevention and promoting animal health and welfare. It is also useful to provide evidence in support of current treatment strategies, help reduce antibiotic use and maintain consumer confidence.[22]

Before taking diagnostic samples there are several considerations:

1. *Is there a requirement to take samples?*
2. *What are the chances of getting meaningful results?*
3. *How will the results influence what is happening on farm?*
4. *Discuss animal and sample selection with the vet to ensure testing is likely to be representative*
5. *Vet to discuss with the laboratory where testing will take place – if there are any doubts about submission, processing or test selection*
6. *Costs involved*

Clinical signs

1. **Posture**
   - Dull / depressed indifferent to other calves
   - Head down
2. **Eyes**
   - Dull, sunken (Fig. 6)
   - Runny discharge
3. **Coat condition**
   - Dry, dull, stary coat (Fig. 8)
4. **Ears**
   - One or both dropped
5. **Nose**
   - Clear or purulent discharge (Fig. 14)
6. **Breathing**
   - Increased rate
   - Shallow or laboured
   - Coughing
7. **Temperature**
   - High >39.5°C
8. **Appetite**
   - Slower, reduced or none

Fig. 14: Nasal discharge

Mixed infections are frequently present in diseased calves, and causative agents can be isolated from apparently healthy animals. Therefore, diagnostic test selection needs to be appropriate to both the disease being looked for and the animal being examined.[23]

- *Sample calves that are representative of the group affected*
- *Sample calves that are in the early stages of disease*
- *DO NOT sample chronically affected calves that have received multiple treatments – it is a waste of time*

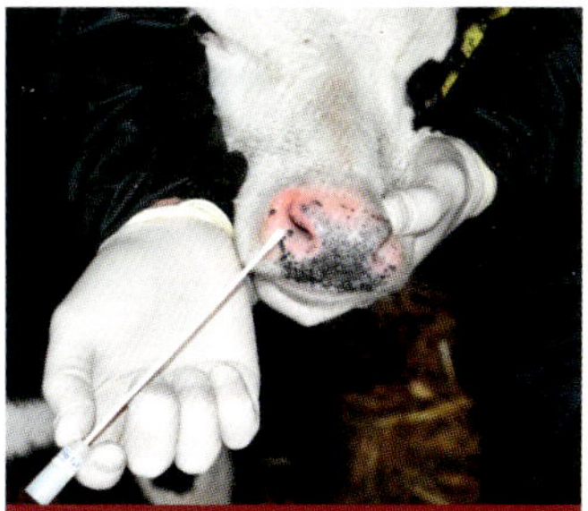

Fig. 15: Performing a nasopharyngeal swab on a calf to test for IBR and bacteriology

Any diagnostic sampling has to be done in consultation with a vet. In practice, the author does the following:

- *Blood sample all calves 1–7 days old on the day of investigation, for colostral status*
- *Blood sample 6 calves aged 4 months or older to assess historical exposure to viruses and mycoplasma on farm*
- *Nasopharyngeal swabs on any suspected cases of IBR (up to 6) and for bacteriology* (Fig. 15)
- *6 x faecal samples in cases of suspected lungworm (see lungworm section page 166)*
- *If a dead untreated calf is available – post-mortem at a suitable laboratory; or send the pluck (trachea and lungs); or histopathology on 3–4 fixed samples of lung tissue that will include "border" regions of healthy and unhealthy tissue*

Treatment

Who do you treat?

"Temp and Treat" is one method to determine which calves to treat. If adequate handling facilities are available, all animals with a temperature greater than 39.5°C should receive treatment.

"Temp and Treat" may be adequate, but scoring systems for calf pneumonia can be more useful. The McGuirk calf health scoring chart (Fig. 16) developed at the University of Wisconsin, Madison, gives a calf's respiratory score by adding the scores from each of rectal temperature, cough, nasal discharge and the highest score from either ocular discharge or ear position. Animals scoring 4 are marked (either individually or their pen) and designated watch, while those scoring 5 or above are treated. Used on a regular basis (ideally twice a week), it allows early recognition and treatment of sick calves. For some farms this does not work though, and "Temp and Treat" is more appropriate, especially in the middle of an outbreak.

In many outbreaks of calf pneumonia there are high disease levels and, on welfare grounds, it is advisable to treat all the calves in the group if multiple calves are affected.[24] The use of antibiotics for prevention (prophylaxis or metaphylaxis) of calf pneumonia must be cautious, due to increasing concerns around the development of antibiotic resistance.

At what point do you initiate metaphylaxis? When faced with an outbreak, treat all animals within a group if more than 10% are showing signs of pneumonia – assessed either by "Temp and Treat" or the McGuirk calf health scoring chart (whichever the farm employs). The justification is that both clinical and subclinical infections receive treatment; it reduces environmental contamination with potentially pathogenic bacteria, and reduces handling of sick, stressed animals who might otherwise exacerbate symptoms. In addition it is likely overall antibiotic usage will be reduced with less repeat treatments.

0	1	2	3
Rectal temperature (°C)			
37.8–38.3	38.3–38.8	38.9–39.3	≥ 39.4
Cough			
None	Induce single cough	Induced repeated coughs or occasional spontaneous cough	Repeated spontaneous coughs
Nasal discharge			
Normal serous discharge	Small amount of unilateral discharge	Bilaterial, cloudy or excessive mucus discharge	Copious bilateral mucopurulent discharge
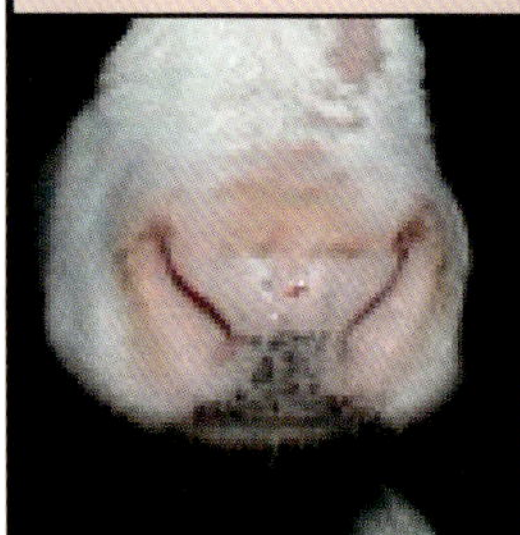		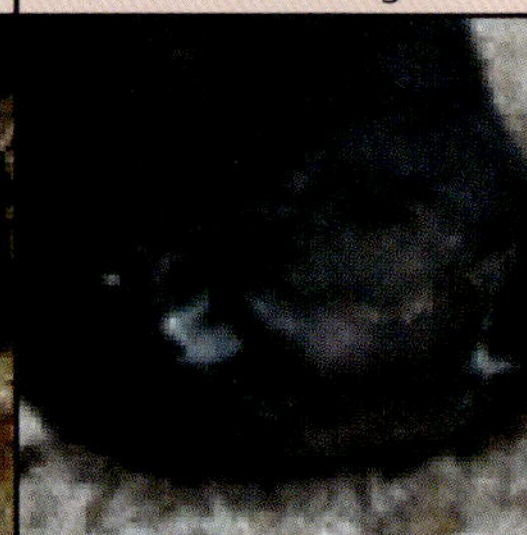	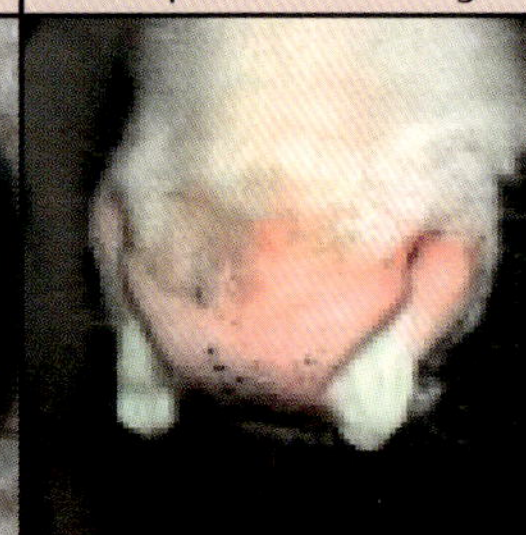
Eye scores			
Normal	Small amount of discharge	Moderate amount of bilateral discharge	Heavy ocular discharge
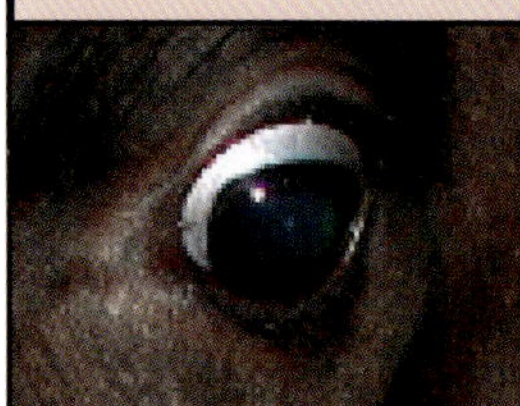	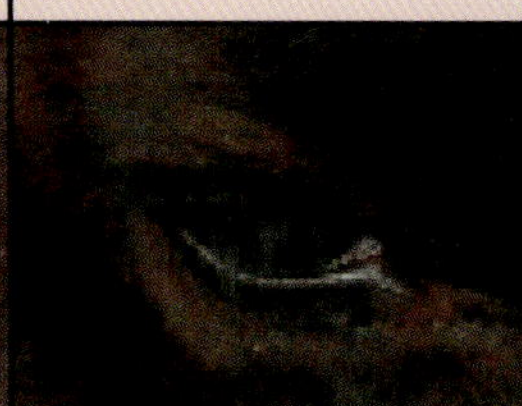	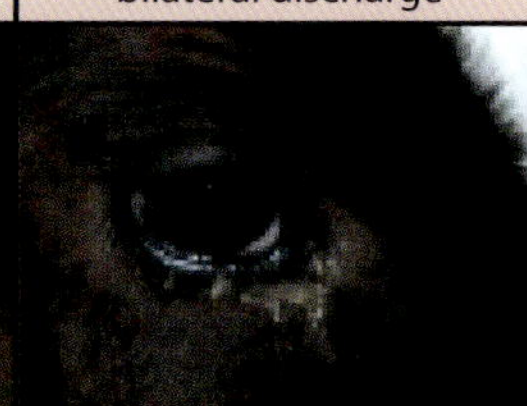	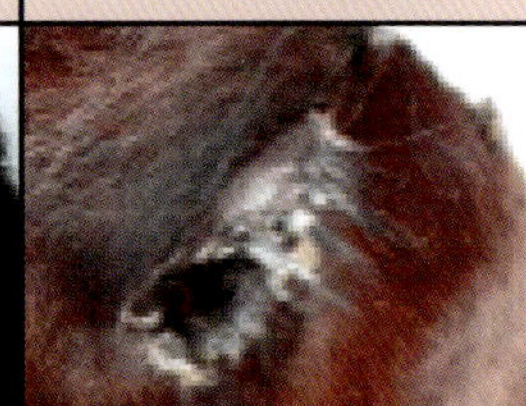
Ear scores			
Normal	Ear flick or head shake	Slight unilateral droop	Head tilt or bilateral droop
		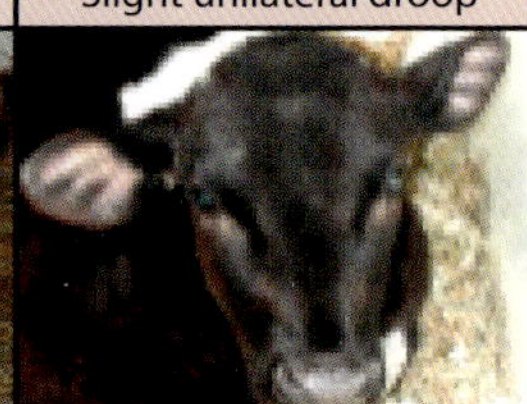	
Fecal scores			
Normal	Semi formed, pasty	Loose, but stays on top of bedding	Watery, sifts through bedding

Fig. 16: The McGuirk calf health scoring chart
(Photo courtesy of S. McGuirk, University of Wisconsin School of Veterinary Medicine[60])

When do you treat?

The single most important factor determining the success of therapy in calves with pneumonia is early onset of treatment, and subsequent adequate duration of treatment.[25] Treatment must be instigated as early on in the disease as possible to reduce the possibility of long-term lung damage and the development of chronic pneumonia.[24] Using the above guidelines of "Who to treat" (page 151) it is imperative that once an animal shows these signs, treatment must be immediate – NEVER wait and see!

What do you treat with?

The specific treatment of calf pneumonia should form part of any farm's herd health plan. The main aim is to cure affected animals. Treatment should be aimed at eradicating the pathogen and any associated inflammatory or hypersensitivity reaction and aiding in the control of fever and depression.[24]

There are no anti-viral drugs available for use in calves so the main component of treatment for calf pneumonia is with antibiotics. There are more antibiotics available for the treatment of calf pneumonia than any other condition in calves, therefore, choice relies on a number of factors. These include previous experience on the farm and elsewhere, reported antibiotic susceptibility patterns, practical considerations such as cost, ease and frequency of use, on-farm handling facilities, and for some products, the issue of human safety. Treatment should always be in consultation with a vet.

In addition to antibiotics, non-steroidal anti-inflammatory drugs (NSAIDs) should be used in the calf with pneumonia. They have several beneficial actions. The anti-inflammatory component reduces lung damage, allowing improved penetration of lung tissue by the antibiotic. They are pain killers – pneumonia is a painful condition, and they reduce fever, which together improves calf demeanour and stimulates feed intake – which will further improve recovery. Their anti-endotoxic actions are also of benefit against certain bacterial infections.

Sick pre-weaned calves may stop feeding so use oral rehydration therapy (ORT) in these cases (see calf scour, page 146).

All animals receiving treatment should be recorded with the date, calf I/D, treatment and quantities administered. This should be done as a matter of course for farm assurance purposes. Ideally it should also be displayed near where calves are housed (Fig. 17). This allows all those dealing with calves to know who has received treatment and monitor progress.

Fig. 17: A white board in the calf shed to record calves treated for disease

Isolate sick calves – either on their own or in groups of sick animals in a separate air space away from other calves. These "hospital pens" reduce the risk of spread of infection to other healthy animals. It also allows closer monitoring of sick calves – assessment of response to treatment, feed and water intake, and additional treatment can be easily administered if required. In practice, isolation is not always possible and, when a whole group of calves is being treated, not necessary.

When do you stop treating?

Three strikes and you're out! If a calf has to receive three treatments for pneumonia it should be culled. This heifer will never keep up with its herd mates, and so remain stunted and have a long delay before entering the milking herd. Heifers suffering more than 4 cases of pneumonia have an average 100 days less productive life.[26] The chronic case also represents a constant source of infection to other animals. Do not waste time on these heifers – sell or cull them.

Control

The most expensive drug in the world is the one that doesn't work. So while vaccination has an important role to play in the prevention of calf pneumonia, good diagnostics are required in the first place to know what the causative agents are. Otherwise incorrect treatment and/or vaccines may be used with poor outcomes.

Vaccines are now available that provide effective control against the most common bacterial, viral and parasitic causes of calf pneumonia. Programmes should be designed to take into account the circumstances on individual farms and should be based around knowledge of the risk factors and the circulating pathogens.[27] Vaccination programmes should reduce disease and death, improve performance, facilitate management and use of resources, have a positive effect on welfare and improve profits.[24]

Basic management issues must be addressed first to ensure that specific pneumonia control strategies have the best possible chance of success, fundamentally, colostrum, nutrition, environment and buildings. A farm visit by a vet that involves all those who have a role in caring for calves is essential for this.

It could be argued that all these other changes should be made first before vaccination is employed. In practice, however, making changes to existing buildings, reducing stocking rates and/or changing feeding patterns, etc. is not always possible straightaway. As much as possible should be done here first. But, the calf sits on one end of a scale, and if vaccination combined with some management changes helps to balance the scale - and stop pneumonia – it can and should be used.

Vaccination is not a magic bullet and can fail in the face of overwhelming pressure from pathogens, so as much as possible should be done to help achieve success:

- *Ensure vaccines are stored and handled correctly on farm*
- *Ensure you know the correct dosage and route of administration for the vaccine (clear labelling)*
- *Check shelf-life and how soon a vaccine must be used once a vial is broached (usually hours)*
- *Use clean equipment in good working order, that is suitable for the vaccine of choice*
- *Vaccinate healthy calves only and separate from stressful events such as transport, mixing, disbudding, etc. by at least a week*
- *Vaccinate all calves in the group*
- *Vaccinate calves at the right age (not too young)*
- *Complete vaccination courses*
- *Dispose of unused product and containers appropriately*
- *Be aware of any adverse reactions and how to respond to them*
- *Be aware of the advice required following accidental self-injection*
- *Read the datasheet for the recommended calf pneumonia vaccine*

Joint and navel ill

Bacterial infection (often *E. coli*) of the navel called "navel ill" or migration of infection to the joints called "joint ill", are associated with poor navel dipping. Navel dipping calves at birth provides many benefits. Without disinfection, the same lifeline that delivers nutrients and removes wastes while the calf is developing in the uterus becomes a dangerous pathway into the calf's body until complete closure takes place throughout the first 2 months of life.

Calves whose navels are not dipped are twice as likely to die as those whose navels are dipped. Those that survive the easily preventable infection are 2.5 kgs lighter at 3 months of age than their herd mates that had their navels dipped at birth.[28]

TARGETS

- 24 hours – 60 days: <1%

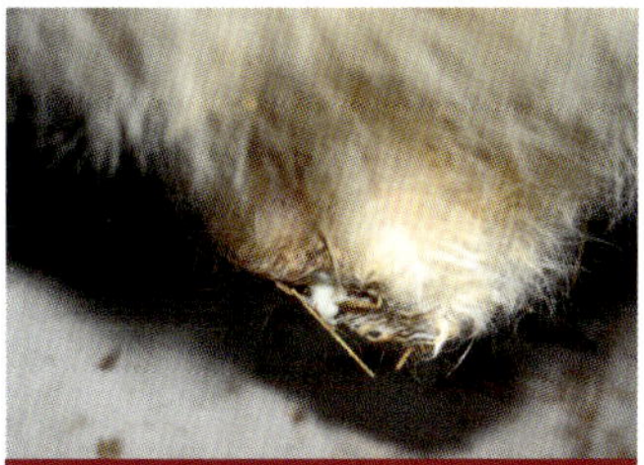

Fig. 18: Navel ill

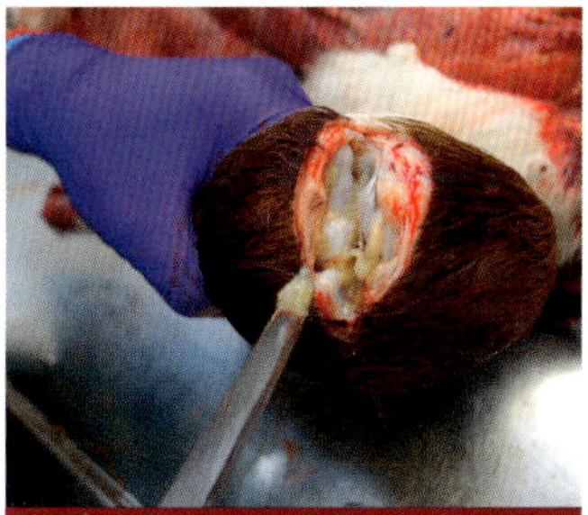

Fig. 19: Joint ill. This joint has been opened up to show the pus that has accumulated within the joint
(Photo courtesy of M. Millar, School of Veterinary Science, University of Bristol)

Clinical signs

Navel ill is usually seen in the calf's first week of life. There may be no generalised signs of illness, but the navel cord feels enlarged and painful, and the tip is usually moist, with a purulent smell (Fig. 18). The temperature will be raised (>39.5^{o}C). More advanced cases will be dull, reluctant to move and pain makes them stand with an arched back. In most cases the infection is limited to the tip of the cord and generally responds well to treatment. However, pus may track up the umbilical arteries, umbilical veins or urachus (which leads to the bladder). Internal abscesses can form, which may even track up to the liver.[29] If infection tracks up the urachus to the bladder it will cause cystitis.

The bacteria may also localise in the joints causing joint ill (Fig. 19). Joint ill is seen later than navel ill at 2–4 weeks and, although infection may have originally entered at the navel, calves with joint ill do not necessarily have an associated navel ill.[29] The first sign is lameness, and swelling around one or multiple joints. If more than one joint is involved, the calf will be lethargic and reluctant to move.[29] It will have a high temperature (>39.5^{o}C). The knees of the front legs and hocks of the back legs are most commonly affected.

Diagnosis

The diagnosis of both navel ill and joint ill is based upon the clinical signs. A hard, swollen, painful, non-reducible umbilicus, represents navel ill (not to be confused with a soft easily reducible, non-painful navel, which is more likely to be a hernia). A swollen joint, with lameness is usually joint ill. If in any doubt consult a vet.

Treatment

Treat navel ill with daily broad-spectrum antibiotics for a week plus NSAIDs daily for 3 days. It is useful to keep the moist end of the cord bathed in warm dilute hibiscrub to encourage pus to drain. If a persistent navel discharge continues there may be a deeper internal abscess which needs to be flushed out – consult a vet.

Joint ill needs early treatment to stand any chance of success. Daily broad-spectrum antibiotics for at least 2 weeks plus NSAIDs daily for 3 days are needed. If a calf becomes recumbent with the condition do not treat – cull.

Control

Dip all calf navels at birth with 10% iodine or chlorhexidine. Make sure the navel is fully submerged or sprayed right up to the skin (Fig. 20). Don't use iodine teat dip – it is not the same thing.

Even those farms with the well-managed calving areas may have trouble keeping them perfectly clean and dry at all times. Navel dipping provides calves with a defence against bacteria in the environment that cannot be revisited once skipped at birth. If navel ill is a persistent problem, repeat dipping at 24 hours of age but, more importantly, check the cleanliness of the calving areas.

The cow and calving environment are the main source of infectious agents to the calf – these include pathogens that cause scour, pneumonia, navel ill and Johne's disease. Remove the calf from the cow immediately after calving. This minimises the risks of infection, and reduces the stress to both the cow and calf.[30] Transport the calf in a clean wheelbarrow or dedicated newborn calf transport vehicle to a clean hutch or pen before administering colostrum.

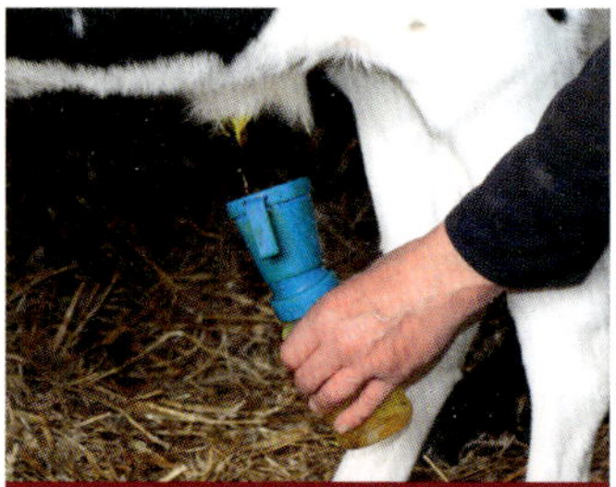

Fig. 20: Dipping a calf's navel with 10% iodine

Ringworm

Infection of hair and skin keratin with the fungi *Trichophyton verrucosum*, and less commonly *Trichophyton mentagrophytes* causes ringworm. This disease is very common worldwide, especially in calves between 2 and 7 months of age, and during the autumn and winter.

TARGETS

- 0 %

Fig. 21: Classic ringworm lesions on two calves.

Fig. 22: More severe ringworm with joined up lesions forming larger areas of infection

Risk factors

Animals kept in close contact with one another are particularly at risk, such as those in intensive management systems. Other species may also be infected, including horses, sheep and also man in whom it may cause serious skin lesions.[31]

Effects on the animal

Ringworm does not give rise to serious debilitating symptoms: its effects are on the value of the animal or its hide, and animals with ringworm may not be shown or sold.[31] If ringworm is present there are often other underlying health problems in a group.

Clinical signs

The spores of ringworm fungi survive for up to 4 years in buildings and on fences. They may be transmitted by carrier animals to susceptible hosts, or transmitted by man, equipment, or simply direct contact with infected buildings and fences.

The incubation period of the disease is approximately 1–4 weeks; the fungus invades hair during its growth phase and uses enzymes to destroy the keratin. The hair then becomes weakened and breaks off, leading to partial alopecia. A generally mild inflammatory reaction usually accompanies the infection.[31]

The disease does not tend to cause itching in cattle. Lesions are greyish-white, with an ash-like surface. Their outline is circular and they are slightly raised due to the accumulation of many layers of scale and the swelling of tissues beneath (Fig. 21). They are typically 30–50mm in diameter, but in more severely infected animals they can join up to form extensive areas of infection[31] (Fig. 22). Distribution is commonly on the head, neck, ears and back.

Ringworm is generally self-limiting with the course of the disease being 1–4 months, but shorter if calves are treated. Re-infection is uncommon.

Diagnosis

Diagnosis is usually made on the appearance of the classic lesions of ringworm. Laboratory examination of hair and crust samples can be performed to identify the fungi, if necessary.

Treatment

Animals often self-cure following turnout in the spring. Topical treatments exist, but are usually reserved for show animals, animals for sale, or severely infected individuals.

Enilconazole (Imaverol, Elanco Animal Health) is the only licensed topical anti-fungal to treat ringworm in cattle. It must be prepared and administered in adequately ventilated surroundings. The concentrated solution should be diluted by adding 1 part to 50 parts of warm water, to provide a 2 mg/ml emulsion (0.2 % w/v).[32] Ringworm will extend into the hair follicles, so possible crusts must be removed with a hard brush, which has been soaked in the diluted emulsion. The removed crusts should be collected and burned to avoid contaminating the premises. It is highly recommended that the animal be sprayed entirely at the first treatment in order to reach the subclinical lesions as well. Depending on the nature of the lesions, cattle should be treated 3–4 times at 3-day intervals. The animals should either be washed with the diluted emulsion or the emulsion should be applied to them with a sprayer or high-pressure cleaning unit.[32]

Control

Thorough cleaning and disinfection of buildings should be performed as soon as they are empty. Sorgene 5 at a 4% solution is effective against ringworm. Halters and grooming equipment can be disinfected with bleach.

Vaccination is available for Ringworm: Bovilis Ringvac (MSD Animal Health). It is available for both the active immunisation of cattle to reduce clinical signs of ringworm (prophylactic dose) and to shorten the recovery time of infected cattle showing clinical signs of ringworm (treatment dose).[33] The duration of protection has not been determined but experience of use in the field suggests that after the recommended course has been completed cattle continue to be protected without re-vaccination.[33]

Internal parasites Nematodes

Cattle can be parasitised by over 18 species of gastrointestinal nematodes – causing infection known as parasitic gastroenteritis (PGE), of which the most economically important species is *Ostertagia ostertagi*. A small intestinal nematode called *Cooperia oncophora*, is commonly found in young cattle in their first grazing season and is often the main contributor to parasite faecal egg counts (FEC).[34] Together, these two gastrointestinal nematodes are of most significance in young cattle, and are focused on in this section. The lungworm, *Dictyocaulus viviparus*, also a nematode, is discussed separately (see page166).

The costs of PGE include treatment, increase in labour to administer the treatment as well as overall animal health and production losses, are estimated to be up to £40 per head.[36]

Ostertagia ostertagi and *Cooperia oncophora* are ubiquitous and found on all cattle farms where there is access to grassland. Clinical disease (loss of appetite, scouring and poor condition) is generally only seen in young calves during their first grazing season, when control has been inadequate. More commonly, infected animals experience loss of production, which is economically important, but difficult to detect without accurate observation and recording. Growth rates can be reduced by 30%, even with a low level of worm challenge, and this makes it difficult to achieve growth targets for replacement heifers.[35]

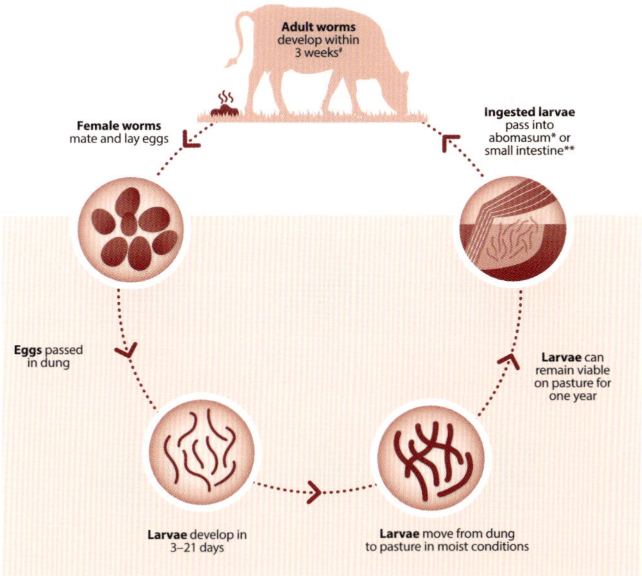

* *Ostertagia* ** *Cooperia* # Development can be inhibited in autumn and resume in late winter or early spring

Fig. 23. *Ostertagia* and *Cooperia* gut worm life cycles

Description

Adult *Ostertagia* are slender reddish-brown worms, males' measure 6–8 mm, and females 8–11 mm in length. Adult *Cooperia* are pinkish-white worms, males' measure 5.5–9 mm, and females 6–8 mm in length.[37]

Life cycle (Fig. 23)

Larvae hatch from worm eggs passed out in faeces. Rate of development is largely temperature-dependent (providing adequate moisture is available in the pat) and can be completed in less than a week when environmental temperatures are 15–23°C and within 3 weeks at 10°C. Moisture enables infective larvae to leave the pat and get on to herbage, where they are more likely to be ingested by cattle. Larvae have limited mobility, but in moist conditions can migrate a few centimetres from faeces. More importantly, rainfall can soften the pat and help to disperse larvae over greater distances.[35]

SITE	SPECIES	FEATURES	DEGREE OF EFFECT ON CATTLE ((H/M/L)
Abomasum			
	Ostertagia ostertagi	Main cause of PGE	**H**
	Ostertagia leptospicularis	Deer parasite. Similar effects to O. ostertagi	**M**
	Haemonchus contortus	Sporadic	**H**
	Trichostrongylus axei	Heavy infections may cause PGE	**M**
Small Intestine			
	Cooperia oncophora	Very common in young cattle in their first grazing season and main contributor to FEC	**L**
	Cooperia pectinata	Uncommon in UK cattle	**M**
	Cooperia punctata	Uncommon in UK cattle	**M**
	Cooperia surnabada	Similar to *C. oncophora*	**L**
	Trichostrongylus colubriformis	Mainly a sheep parasite	**L**
	Trichostrongylus longispicularis	Uncommon in UK cattle	**L**
	Nematodirus helvetianus	Occasional and potentially causes problems in cattle	**M**
	Bunostomum phlebotomum	Uncommon in UK cattle. Can cause anaemia	**M**
	Strongyloides papillosus	Relatively common but only very young calves may rarely show signs of disease	**L**
	Toxocara vitulorum	Rare	**L**
	Capillaria bovis	Common but little significance	**L**
Large intestine			
	Oesophagostomum radiatum	Common but little significance	**M**
	Trichuris globulosa	Common but little significance	**L**
Lungs			
	Dictyocaulus viviparus	Highly problematic; increasingly reported and distributed	**H**

Table 3: Nematode parasites of cattle (Adapted from Taylor[34])

Once on pasture, infective third-stage larvae (L3), are relatively resilient and can survive for many months and in small numbers over a year under normal environmental conditions in the UK. The L3 are sensitive to desiccation, so survival is reduced during hot, dry spells of weather.[35]

Following ingestion, infective larvae pass to their preferential sites for development – the abomasum for *Ostertagia* and upper small intestine for *Cooperia*. Larvae establish and develop into adult worms in about 3 weeks, and then females start to lay eggs.

Inhibition of larval *Ostertagia* is an adaptation for survival over winter, but can also result in disease. Usually only a few cattle are affected. Typically when cattle are grazing in autumn, a high proportion of larvae ingested stop developing a few days after entering the abomasum. They remain dormant within the gastric glands causing very little damage but, in late winter or early spring, something triggers them to resume development. If the numbers of developing larvae are small, consequences are insignificant, but if large numbers of larvae resume development and emerge as adults simultaneously – there is considerable damage to the abomasum with severe clinical signs, even death.[35]

Pattern of disease

Over winter, due to lower temperatures, there is limited and slow development of nematode eggs and larvae in dung pats that were deposited before housing. A decline in numbers of infective larvae occurs on pasture, as these stages do not feed and slowly use up their energy reserves. Larvae numbers actually decline more rapidly during a mild winter – because higher temperatures mean metabolism is faster and energy reserves disappear at a faster rate. Following a cold winter, particularly under snow, survival is higher as larvae exist in a dormant state at low temperatures and snow cover can prevent dehydration.[35]

A considerable number of L3 larvae can survive the winter on pasture and in soil. Sometimes, numbers are sufficient to cause disease in weaned calves 3–4 weeks after they are turned out to graze in spring (Type I disease). This is unusual, however and the role of the surviving L3 is rather to infect calves at a level, which produces sub-clinical infection and ensures contamination of the pasture for the rest of the grazing season.[37]

The loss of most overwintered L3 on the pasture occurs in spring and only negligible numbers can be detected in June. This loss along with the dilution effect of rapidly growing herbage renders most un-grazed pastures in the spring safe for grazing after mid-summer. However, some L3 may survive in the soil for at least another year and can subsequently migrate onto the herbage.[37]

Once mature, adult female worms are prolific egg layers and infected calves pass high concentrations of eggs in dung. A new generation of eggs develops into larvae within dung, but as temperatures are typically low in spring, this takes several weeks to complete. As the season advances and temperatures increase, development time to infective larvae shortens and, if calves remain on the same pasture and are not treated, larval numbers on pasture typically peak in early July and remain high until animals are housed. These high levels lead to outbreaks of clinical disease.[35]

As autumn progresses temperatures decline and an increasing proportion (up to 80%) of the L3 ingested become inhibited at the early fourth larval stage (EL4). In late autumn calves can therefore harbour many thousands of these EL4 but few developing forms or adults. There are no symptoms until maturation of EL4 takes place during winter and early spring when Type II disease may occur. Where maturation is not synchronous, clinical signs may not occur, but the adult worm burdens that develop can play a role in contributing to pasture contamination in the spring.[37]

The prevalence of Type II ostertagiosis in dairy calves is increased by:

- *Grazing from May until late July on permanent pasture, then moving calves to hay or silage aftermath for 2–3 months before returning them to the original grazing in late autumn. The permanent pasture will contain many L3 and when ingested they will become arrested*[37]
- *In dry summers L3 are retained within the crusted faecal pat and cannot migrate onto the pasture until sufficient rainfall occurs. If rainfall is delayed until late autumn many of the larvae liberated onto pasture will become arrested following ingestion, increasing the chance of Type II disease.*[37]

Significant immunity does not develop in the calf until the end of the first grazing season, and this then wanes over the winter housing period. Yearlings turned out the following spring are therefore partially susceptible to re-infection and contaminate the pasture with small numbers of eggs. Immunity is then quickly re-established and any clinical signs are brief. By the second and third year of grazing, adult stock in endemic areas are usually highly immune to re-infection and won't affect the pattern of disease. Around calving though immunity can wane again in heifers and clinical disease is occasionally seen after calving.[37]

Effects on the animal

Large populations of *Ostertagia* can cause extensive damage and biochemical changes, which peaks when the parasites emerge from the gastric glands lining the abomasum. In heavy infections of 40,000+ adult worms the effects are:

1. *Reduced acidity in the abomasum from 2.0 to 7.0. This prevents activation of an enzyme (pepsinogen) responsible for protein digestion, and reduces the bacteriostatic effects in the abomasum*
2. *Large molecules can leak out of the abomasum. Combined this leads to loss of enzymes and lowered protein levels – which can become serious*

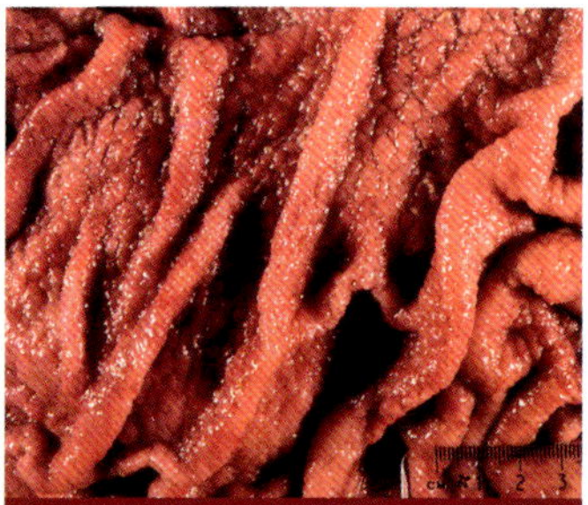

Fig. 24: Adult *Ostertagia* worms can be seen on close inspection of the abomasal surface. (Reproduced with permission from Diseases of Cattle in Australasia, published by VetLearn, Wellington, New Zealand)

Clinical signs

1. *Profuse watery diarrhoea*
2. *In Type I disease, where calves are at grass, this is persistent and bright green*
3. *In Type II disease diarrhoea is often intermittent and anorexia and thirst are usually present*
4. *In both forms of disease, loss of body weight is considerable, reaching up to 20% in 7–10 days*
5. *Cooperia oncophora is generally considered to be a mild pathogen of calves but has sometimes been associated with loss of appetite, poor weight gain and intermittent diarrhoea.*

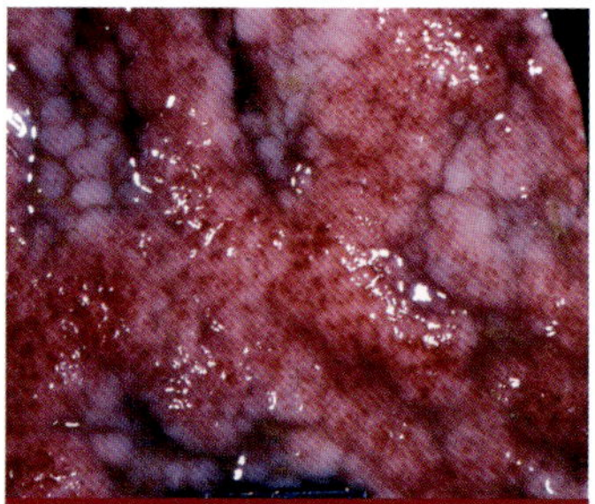

Fig. 25: The lining of the abomasum has become thickened and nodular. (Reproduced with permission from Diseases of Cattle in Australasia, published by VetLearn, Wellington, New Zealand)

Diagnosis

- *Clinical signs of inappetance, weight loss and diarrhoea*
- *Season:*
  - *Type I = July–September*
  - *Type II = March–May*
- *Grazing history:*
  - *Type I = calves have usually been set-stocked in one area for several months*
  - *Type II = calves grazed on one field from spring to midsummer, then moved and brought back to the original field in autumn*
  - *There is also usually a history of Ostertagiosis on a farm*
- *Faecal egg count (FEC).*
  - *In Type I disease >1000 eggs per gram (epg) is a useful aid to diagnosis.*
  - *However, it does not give a direct correlation to*
    (i) The current population of adult and/or immature worms in the gastrointestinal tract of individuals;
    (ii) The damage caused by the worms or knock on effects;
    (iii) Growth rate (as affected by worms) in young cattle
- *For these reasons FECs must never be used in isolation but along with other means of diagnosis*
  - *In Type II disease FEC is highly variable and of limited value*
- *Plasma pepsinogen levels. In clinically affected animals up to 2 years old these are usually in excess of 3.0 iu tyrosine (normal = 1.0 in non-parasitised calves)*
- *Post-mortem examination – adult worms can be seen on close inspection of the abomasal surface (Fig. 24). Over time the lining of the abomasum itself changes to become thickened and nodular (Fig. 25).*

Treatment

Anthelmintics (wormers) are used to treat nematodes. There are five groups:

1. *Benzimidazoles (BZ)*
2. *Levamisole (LV) (Imidazothiazoles)*
3. *Macrocyclic lactones (ML) including avermectins and milbemycins*
4. *Amino acetonitrile derivatives (AD) (Monepantel)*
5. *Spiro-indoles (SI) (Derquantel available as a multi-active)*

These groups are active against the major species of nematodes. Some also have activity against liver fluke and tapeworms. MLs injectables and pour-ons also have activity against some external parasites.

Choosing the most appropriate product for the parasites likely, or known, to be present is vital. Targeting the right parasite gives predictable results and reduces the need for re-treatment. It also reduces selection pressure for anthelmintic resistance.[38] A discussion with a vet should be undertaken to assess risk and the most appropriate treatments. Administer anthelmintics effectively (Table 4).

ADMINISTERING ANTHELMINTICS EFFECTIVELY
Choose the most appropriate product for the parasites likely, or known, to be present
Store wormers in accordance with instructions – away from direct sunlight, extremes of temperature and sometimes refrigerated
Always read the label before using any product to ensure it is suitable for the livestock to be treated and be aware of any precautions for use
Use products before they expire
Make sure dosing equipment is compatible with the product and check it is clean and measuring the correct volume
Administer products according to manufacturer's instructions, paying particular attention to specific methods for ear injections and intra-ruminal boluses
Dose according to liveweight – use a scale or weighband
Record all products administered in the medicines book
Note any withdrawal periods and adhere to them
Do not mix wormers with any other products – it can inactivate them

Table 4: Administer anthelmintics effectively

To treat nematodes, anthelmintics can be used:

1. **Strategically** – to target groups of animals and limit pasture contamination with worm eggs
2. **Therapeutically** – to treat individuals or groups that are already suffering losses.[35]

1. Apply strategic treatments early in the grazing season – at or within 3 weeks of turnout (the time it takes for adult worms to develop) because this ensures worm eggs do not add to pasture contamination. Then aim to minimise egg contamination of pasture up to mid-July, by which time the over-wintered larval population should decline to insignificant levels.
Options include administration of long-acting injection or bolus formulations at turnout or administration of macrocyclic lactones (MLs) that have a shorter persistence (3–6 weeks) at turnout and then again after an interval of 6–8 weeks; occasionally a third treatment after the same interval is required for a long grazing season.[35]
This approach assumes cattle remain on the same pasture throughout the grazing season. If they are moved, then the risk posed by the next pasture needs to be assessed and steps taken to maintain control.[35]

2. If measures haven't been taken to limit pasture contamination with parasites through pasture management or anthelmintics, grazing cattle will be exposed to ever-increasing risks of disease over the season and, consequently, they will experience production losses and suffer clinical PGE.

   Type I disease responds well to treatment at the standard dosage rates with any of the modern benzimidazoles, levamisole or the avermectins/milbemycins. All are effective against developing larvae and adult worms. Following treatment, move calves to pasture that has not been grazed by cattle in the same year.

   Type II disease requires modern benzimidazoles or avermectins/milbemycins as they are all effective against arrested (hypobiotic) larvae, developing larva and adult worms.

Animal performance is superior in young cattle that are not exposed to high levels of challenge (i.e. in strategic programmes), so adopt this approach.

A housing treatment can be considered as strategic, therapeutic or both: therapeutic because worm burdens are removed and cattle are free of their negative impact for the remainder of the housing period: and strategic because when cattle go out to pasture the following year, they will not immediately contaminate pasture with worm eggs.[35] In addition with appropriate products (see above) Type II disease is minimised.

Use more than one parasiticide or a combination product at housing, as this will also treat lungworm, liver fluke, and prevent lice and mange infestations later in winter.

Anthelmintic resistance is a serious issue in sheep and will inevitably become so for cattle. Best practice when using these products is essential to prolong the time they are useful for. Table 5 lists best practice guidelines to limit the development of anthelmintic resistance.

GUIDELINE		COMMENT
1.	Work out a control strategy with your vet	This should form part of an annual herd health plan and will require ongoing consultation
2.	Use effective quarantine strategies to prevent the importation of resistant worms in introduced cattle	Bought in cattle can be a potential route of introducing resistance into a non-closed herd
3.	Test for anthelmintic effectiveness on your farm	While resistance is still rare in cattle nematodes, treatment failures do occur. It is important to monitor continued effectiveness as underdosing can select for anthelmintic resistance
4.	Administer anthelmintics effectively	Administer the right dose in the correct way by following the manufacturer's instructions
5.	Use anthelmintics only when necessary	Understand the trade off between tolerating some level of parasitism and minimising selection for anthelmintic resistance. FEC monitoring has an important role
6.	Select the appropriate anthelmintic for the task	Target treatment according to parasites (and their stages) present, based on time of year
7.	Adopt strategies to preserve susceptible worms on the farm	Aim to reduce selection for anthelmintic resistance when treating adult cattle, immune older animals or when dosing on low contamination pastures
8.	Reduce dependence on anthelmintics	Alternative control measures include grazing management using sheep or older immune animals

Table 5: Guidelines for the control of nematodes and use of anthelmintics in cattle (Adapted from COWS[34])

Control

Anthelmintics have a key role to play in nematode control, but an integrated approach, bespoke to each farm, is required to manage the disease. It is important to consider:

- *Production objectives, i.e. growth rates and age at first calving*
- *Farm infrastructure, particularly in relation to pastures, grazing management and handling facilities*
- *Presence of other parasites, such as lungworm and liver fluke*
- *Effectiveness of available anthelmintics.*

A risk assessment for PGE can then be compiled (Table 6). Review it several times during the grazing season. Housing and turnout are also good times to reflect on the previous months and plan for the future. Once risks have been quantified, it should be possible to decide on control options, using management and anthelmintics. Choice of anthelmintic should be done in consultation with a vet.

RISK FACTOR	HIGH	MEDIUM	LOW
Age (grazing seasons)	<1 year (first grazing season)*	1–2 years (second grazing season)	>2 years (adult)*
Age at turnout for weaned calves (first grazing season)	<6 months	6–8 months	>8 months
Weight gain (<2 years old) two months after turnout	<0.7 kg/day	0.7–0.8 kg/day	>0.8 kg/day
FEC two months after turnout	>200	50–200	<50
Herbage mass, Kg DM/ha Sward height, cm	<1000 <4	1000–2000 4–8	>2000 >8
Field type	Permanent pasture	Silage/hay aftermath	Newly sown, ungrazed leys
Grazing history	Grazed by cattle <1 year old within the last year	Grazed by cattle 1–2 years old within last year	Grazed by adult cows, sheep** or other species within last year
Bulk milk tank *Ostertagia* antibodies	>0.8	0.5–0.8	<0.5
Snail habitats	Widespread	Patchy	Isolated/none
Faecal fluke egg count	+++ (>20)	+ (1–20)	0
Liver fluke diagnosed	Previous year	>5 years previously	Never***
Lungworm diagnosed	Previous year	>5 years previously	Never***

*Adult cattle rarely suffer from clinical PGE, but are susceptible to lungworm (if immunity is low) and liver fluke
**If sheep are infected with liver fluke, they can increase the risk of liver fluke in cattle
***Beware of bought-in animals that could bring in infection (fluke and/or lungworm) or be immunologically naive (lungworm)

Table 6: Grid for assessing risk of disease and production losses from parasitic nematodes in cattle (Adapted from COWS[39])

Give the most susceptible (first and second season) cattle preferential access to low-risk pastures (such as silage aftermaths). Develop a pasture plan for the whole farm mapped for the whole season, noting when low nematode risk aftermaths become available, and produce a grazing plan covering all stock (cattle and sheep). Base the rotation of stock through the pastures on susceptibility and pasture risk (Table 6)

Assessing growth rate regularly throughout the grazing season is the key objective measure to assess if there is a problem with PGE. The target is 0.8 kg/day for heifers to achieve an AFC of 24 months. Weigh heifers at turnout and every month throughout the grazing season.

Lungworm

Lungworm also known as husk, hoose, verminous pneumonia, dictyocaulosis and parasitic bronchitis is caused by the nematode *Dictyocaulus viviparus*. Economic losses are associated with reduced growth rates, treatment, veterinary costs, lost milk revenue, reduced fertility performance and mortality. In severe lungworm outbreaks in growing cattle losses can average £50–£100 per head.[40]

Fig. 26 shows the number of diagnoses of lungworm made each year by AHVLA over the 10 years to 2013. These figures need to be looked at as a percentage of diagnosable submissions to get any idea of trends, as total submissions vary over the years (Fig. 27). The general trend is down from above 5% as a percentage of diagnosable submissions in 2004 to 3% in 2013.

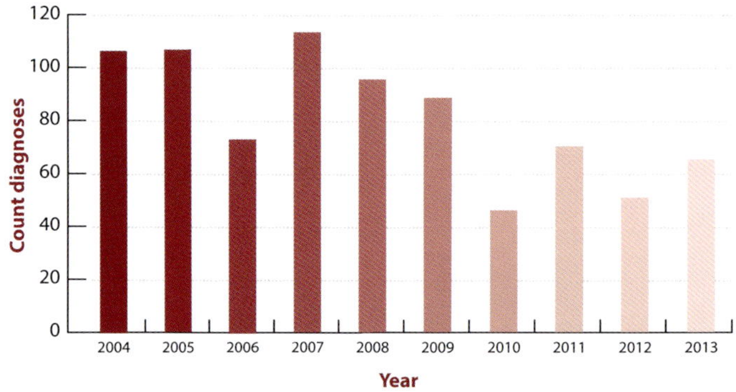

Fig. 26: Diagnoses of lungworm made each year by AHVLA

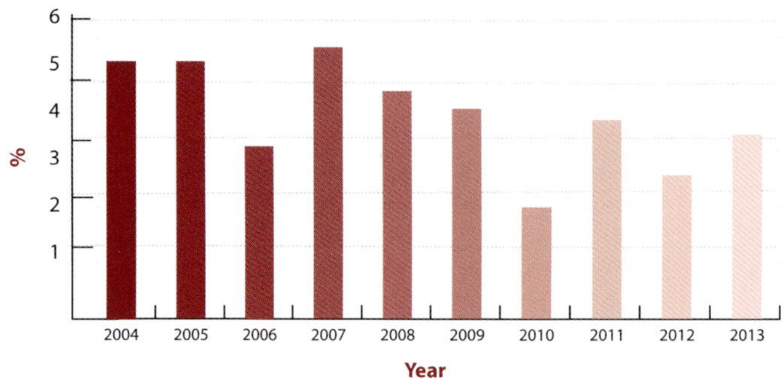

Fig. 27: Incidence of lungworm in cattle as a percentage of diagnosable submissions (AHVLA submissions)

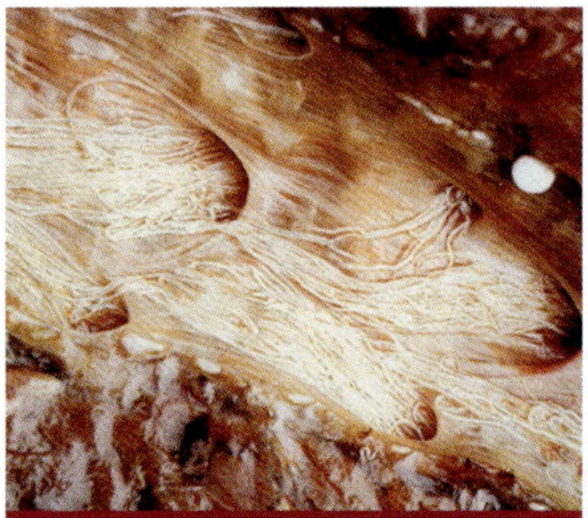

Fig. 28: Adult lungworm. (Photo courtesy of MSDAH)

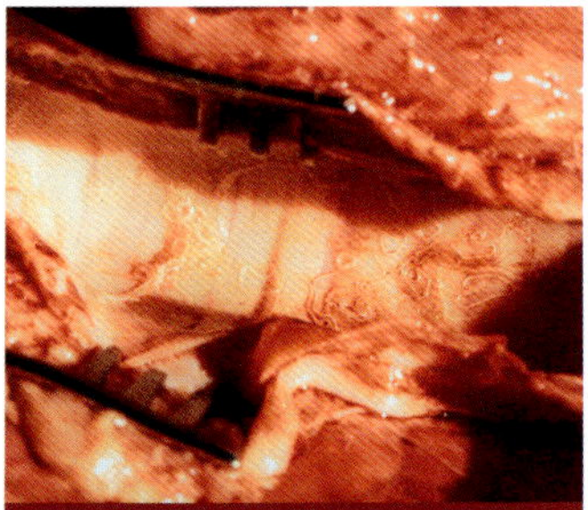

Fig. 29: Adult lungworm within the trachea of a dead heifer. (Photo courtesy of MSDAH)

Fig. 30: Pilobolus fungus growing on a faecal pat. (Photo courtesy of MSDAH)

Description

Adults are slender thread-like worms: males measure 4.0–5.5 cm and females are 6–8 cm in length[37] (Fig. 28).

Hosts

Lungworm occurs worldwide, predominantly in areas with a warm, wet climate for at least part of the year. In Great Britain the disease is most common in the wetter, western parts of the country.[41] Hosts include cattle, buffalo, deer and camel.[37]

Life cycle

Animals can harbour thousands of worms in the lungs (Fig. 29), and adult females produce several thousand eggs in a lifetime. Eggs contain fully developed larvae that hatch almost immediately.[34] Larvae migrate up the trachea or are coughed up and then swallowed in mucus, finally passing out in faeces.

Development to infective L3 larvae takes place within the faecal pat. L3 larvae can develop within 7 days of excretion in faeces, so under appropriate environmental conditions, levels of larval challenge build up quickly.[42] L3 larvae leave the faecal pat to reach herbage by their movements or windborne spread utilising *Pilobolus* fungus, which grows on faecal pats[34] (Fig. 30).

After ingestion L3 larvae penetrate the intestinal lining and develop and pass via the lymphatics and blood to the lungs – they finally break out of the capillaries into the alveoli about 1 week after infection. The final moult occurs in the lower airways a few days later and young adults then move up the bronchi and mature.[34] This takes about 25 days.

Pattern of disease

D. viviparus differs from other cattle nematodes in a number of important respects.

- *Female worms produce very large numbers of larvae*
- *Infective dose required to cause disease is smaller*
- *Resistance to infection is more rapidly acquired*
- *Migration of larvae is more dependent upon external and physical climatic factors than other infections*[41].

These factors make lungworm a very unpredictable disease, both in timing and severity. Outbreaks of disease occur from June until November, but most commonly from July to September. Important factors in predisposing to an outbreak of clinical lungworm disease include:

- *A history of D. viviparus on a farm as evidenced by previous outbreaks*
- *Levels of moisture affect translation of L3 larvae onto pasture as not only does rain accelerate larval dispersion from pats but also in the warmth of summer, moist conditions encourage the growth of Pilobolus helping disperse L3 larvae. A dry summer one year followed by a wet summer the next increases the probability of disease outbreaks. Dry weather results in low translation of L3 larvae onto pasture in one year, creating a natural "immunity gap" in grazing animals leaving them susceptible to disease in the subsequent grazing season*
- *Introducing susceptible stock can destabilize immunity in a herd where a low level of L3 larvae challenge, insufficient to cause disease in existing stock previously, can be sufficient to instigate an outbreak in the introduced animals, thus amplifying levels of infection on pasture.*[42]

Infection can persist in endemic areas in two ways. First, overwintered L3 larvae survive on pasture from autumn until late spring in sufficient numbers to initiate infection or occasionally cause disease. Secondly, carrier animals, particularly yearlings, can harbour small numbers of adult worms in their bronchi until the next grazing season.[37] Lungworm can also be introduced to a herd from a neighbouring premises, by airborne spread (when the spores of a *Pilobolus* fungi are discharged they are projected up to a distance of 3m in still air to land on surrounding herbage[37]) or by mechanical transmission from infected faeces.[41]

Lungworm is most often seen in dairy or cross-dairy calves. Traditionally, only calves in their first grazing season were clinically affected, since on farms where disease is endemic older animals have strong acquired immunity.

Following the introduction of widespread vaccination in the early 1960s, reports of disease in adult cattle were rare but, since 1993, parasitic bronchitis has become the most common diagnosed respiratory disease of adult cattle according to VIDA records.[41,42] Almost three-quarters of cases in 1999 occurred in second season or older animals, and the adult proportion was the highest yet recorded to that date.[43] A move away from vaccination to long-acting anthelmintics has meant young animals do not develop immunity to lungworm as they are not exposed to their larvae. When animals are turned out for a second season without being wormed, exposure to lungworm larvae results in clinical disease in second season or older animals.

Effects on the animal

The effects can be divided into four phases:

- *Penetration phase (days 1–7): larvae make their way to the lungs.*
- *Prepatent phase (days 8–25): appearance of larvae within alveoli where they cause alveolitis, followed by bronchiolitis and bronchitis as they become immature adults and move up the bronchi. Accumulation of white blood cells, inflammatory tissues and cellular debris causes collapse of alveoli, which is largely responsible for the first clinical signs. Heavily infected animals, whose lungs contain several thousand developing worms, may die from day 15 onwards due to respiratory failure.*[37]
- *Patent phase (days 26–60): a parasitic bronchitis characterised by hundreds or even thousands of adult worms in the frothy white mucus in bronchi. In addition, there are dark red collapsed areas around infected bronchi – a parasitic pneumonia caused by the aspiration of eggs and L1 larvae into alveoli and subsequent inflammation.*[37]
- *Post-patent phase (days 61–90 in untreated calves): the recovery phase after adult lungworms have been expelled. Clinical signs abate although the bronchi are still inflamed, with the system eventually becoming normal. However, in about 25% of animals, which have been heavily infected, there is a flare-up of clinical signs, which is frequently fatal. Calves may also die in the post-patent phase if brought indoors and they suffer a superimposed bacterial infection leading to acute pneumonia.*[37]

Clinical signs

Coughing and an increased rate and depth of breathing in any age group of cattle at grass should trigger a suspicion of lungworm.[41] Within any affected group, differing degrees of severity are apparent.

- *Mildly affected animals cough intermittently, particularly when exercised.*
- *Moderately affected animals have frequent bouts of coughing at rest, with faster, deeper breathing (respiratory rate >60).*
- *Severely affected animals show very fast breathing (respiratory rate >80) and difficulty breathing and frequently adopt the "air-hunger" position of mouth breathing with head and neck outstretched* (Fig. 31). *A deep harsh cough, salivation, anorexia and sometimes fever are present.*[37]
- *All affected calves will show a reduced weight gain.*

It is not unusual for all calves in a group to be affected, with death varying with size of infective dose and speed at which treatment is instigated. In approximately one in three outbreaks at least one death can be expected, rising up to 14% in severe outbreaks.[44]

Diagnosis

Clinical signs, season and grazing history will suggest a diagnosis of lungworm. Once mature parasites are in main respiratory passages, larvae can be detected in faeces.[45]

Faecal samples can be submitted for analysis of larvae – a minimum of 15 g faeces is required. False negative results can occur if infection is still in the prepatent period – the animal can be showing clinical signs up to 2 weeks before larvae are detected. When investigating an outbreak, analyse faecal samples from a group of 6–10 animals that have been showing clinical signs the longest.

Treatment

All available anthelmintics are highly effective against developing larvae and adult *D. viviparus*.[37] Early treatment reduces the damage and all individuals in a group should be treated. Early treatment may mean animals are not immune to further re-infection and disease, therefore, do not return animals to the field that was the source of infection.[45] House and offer supportive treatment to severely affected individuals in the form of re-hydration, NSAIDs and antibiotics if secondary bacterial infection is suspected.

Fig. 31: A calf severely affected with lungworm adopting the "air-hunger" position of mouth breathing with head and neck outstretched (Reproduced with permission from Diseases of Cattle in Australasia, published by VetLearn, Wellington, New Zealand)

There are some specific considerations of treatment:

1. *In severe well-established cases in a number of calves, anthelmintic treatment, while being the only course available, may exacerbate clinical signs in one or more animals with death in some cases.*[37]
2. *There are possible disadvantages with the use of macrocyclic lactones due to mass worm death and possible anaphylactic shock in cattle with heavy worm burdens. There is a lack of evidence-base, but the use of corticosteroids concurrently with anthelmintics may reduce this risk of anaphylaxis. Levamisole-based products are thought to be advantageous in that they paralyse the worms rather than killing them – avoiding leaving a lung full of dead parasites. It is then hoped natural defence mechanisms will clear infection from the lungs.*
3. *Consideration of milk withdrawal times should be taken when treating animals that may enter the milking herd shortly.*

Control

Control of lungworm must be planned alongside control strategies for other gastrointestinal nematodes and take into account grazing plans, growth targets and other farm management policies.[41] Control by using grazing systems that have been successful in preventing PGE in cattle has seldom been effective in controlling lungworm due to the unpredictable pattern of disease.[45] Similarly long-acting/continuous-release anthelmintic treatments may protect calves in their first grazing season but do not address the problem of lifetime immunity.[41] The most effective way of controlling lungworm is vaccination.

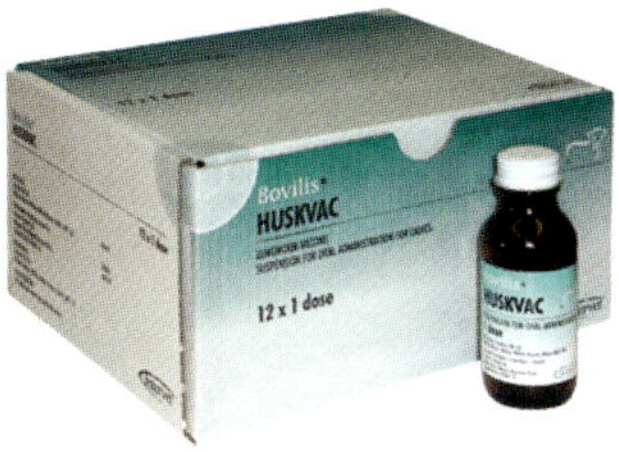

Fig. 32. Bovilis Huskvac

Bovilis Huskvac (MSD Animal Health) is the only licensed product in the UK for vaccination of cattle against lungworm (Fig. 32). Each single dose (25 ml) contains 1000–2000 viable *D. viviparus* 3rd stage irradiated larvae. Healthy animals 8 weeks or older can be vaccinated (Table 7). Irradiated larvae reach the lungs without inducing clinical signs, but develop sufficiently to stimulate immune protection against further challenge.[45]

Basic vaccination scheme:
Two doses at a dosage interval of approximately 4 weeks.

Re-vaccination:
Lungworm immunity is maintained from season to season by the exposure to lungworm larvae, which in most cases occurs from the grazing of normal pastures after vaccination. Under these conditions of exposure, re-vaccination is generally not required.

A single dose of Bovilis Huskvac prior to each season's turnout will boost immunity where such exposure has not occurred, e.g. extensive use of anthelmintics or if using reserved or clean pasture for a large part of the grazing season

Table 7. Vaccination regime For Bovilis Huskvac (MSD Animal Health)

Vaccination provides a buffer against variable exposure to natural infection required to maintain immune status within a herd. Vaccination in the first year at grass will also provide an initial immune stimulus to animals receiving long-acting anthelmintic preparations that may not allow exposure to field infections. Vaccination does not guarantee absence from disease, as exposure to field infection is required to maintain immunity.[41]

Cattle must not receive an anthelmintic within 2 weeks of the second dose of Bovilis Huskvac, or irradiated larvae are killed before they can generate a suitable immune response. The number of anthelmintic products and their variation in action and duration make timing of vaccine administration crucial (see Table 8).

Worming strategy	Vaccination strategy
Herds that do not use any anthelmintic cover in their first season, or use dose and move or strategic weekly administration of long-acting injectables or pour-ons	Administer a full course of vaccine prior to turnout. Do not turn out onto clean pasture but onto pasture where there is an opportunity for exposure to natural infection. For example, pasture grazed by adult cattle last season
Herds using long-acting/continuous release anthelmintics (including fenbendazole slow release boluses) that give cover throughout the ENTIRE first grazing season	Administer a full course of vaccine prior to turnout in the second season
Herds using long-acting/continuous release anthelmintics (including fenbendazole slow release boluses), but where calves are to remain at pasture after expiry of the bolus or treatment regimen	Give a full course of vaccination prior to first season (with second dose at least 2 weeks before worming). In areas of high lungworm risk give single booster dose prior to turnout in second season
Herds that use long-acting/continuous release anthelmintics in the first AND second season at grass	Seriously review such a policy, for there is often little indication for disease control and it can seriously compromise lungworm immunity. If this is not possible, consider a full course of vaccine prior to entry into the milking herd
Herds in which replacements are only turned out in their second season	Administer a full course of vaccine prior to turnout

Note that when choosing a worming strategy for use after vaccination, a product that allows "leakage" or gaps in its cover is to be preferred. This is because it allows natural challenge to boost the level of immunity already gained from the vaccine. Note, however, that pulsed wormers only permit challenge if worms happen to be prevalent during the gaps in their activity.

Table 8. Anthelmintics and their impact on vaccination

It is important lungworm vaccination is continued each year in all new groups of calves. On average, 63% of farms that cease vaccination have an outbreak in succeeding years.[43]

Liver Fluke

The liver fluke *Fasciola hepatica* is very common in cattle in the United Kingdom. Fig. 33 shows the incidence of liver fluke as a percentage of diagnosable submissions to the AHVLA nationally, for 10 years up to 2013. Despite the fact that most farmers will treat liver fluke themselves without the need for a laboratory diagnosis, there is a clear trend of increasing liver fluke in the UK.

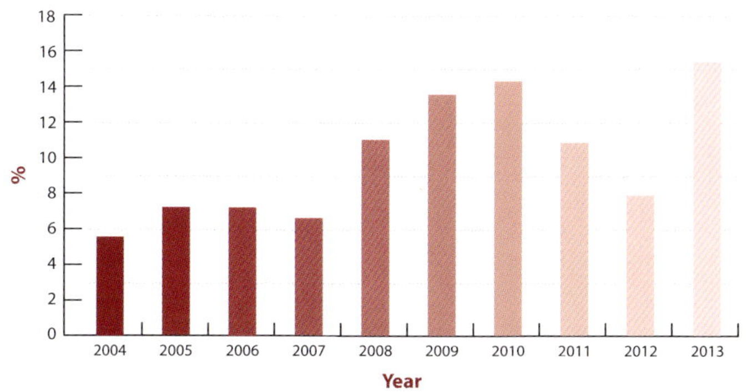

Fig. 33: Incidence of liver fluke as a percentage of diagnosable submissions to the AHVLA

Figure 34: An adult liver fluke

Fig. 35: The mud snail *Galba truncatula*

Fig. 36: A drainage channel – a temporary habitat for *Galba truncatula*

Liver fluke has major impacts on animal welfare and farm profitability. Average losses are £235 per infected animal.[46] Most of these losses are from reduced milk yield and reduced fertility, with smaller losses due to reduced meat production and condemnation of livers.

Description

On entering the liver the young fluke is 1.0–2.0 mm in length and lancet-like. When fully mature in the bile ducts, it is leaf-shaped, grey-brown in colour, 2.5–3.5 cm long and 1.0 cm wide (Fig. 34).

Hosts

Liver fluke occurs worldwide. It can infect all grazing animals, most significantly cattle and sheep but also goats, horses, donkeys, pigs, deer, rabbits and hares. Liver fluke can also infect man. The intermediate host in the UK is the mud snail Galba truncatula (Fig. 35) – a 3–10 mm long brown snail found in permanent boggy areas and in temporary habitats such as hoof prints, drainage channels and flooded areas (Fig. 36).

Life Cycle

The liver fluke has a complicated life cycle that has implications for increase in disease occurrence, diagnostics, treatment options and herd management.[47]

Liver fluke has a two-host life cycle, involving cattle and mud snails. Cattle become infected by ingesting encysted larvae on herbage. Once in the small intestine, immature flukes are released, they then penetrate the intestinal wall and migrate to the liver in about 24 hours.[48]

On reaching the liver, young flukes begin a period of voracious eating and slow movement through the liver tissue towards the small bile ducts.[49] Maturation to adults occurs within the larger bile ducts and gall bladder. Adult flukes can lay up to 5000 eggs/day from about 10–12 weeks post-infection. Eggs are shed into the intestine through the common bile duct and passed in faeces.[48]

In water eggs become embryonated and free-swimming larvae hatch. This only occurs above 10°C. At 15°C hatching occurs 6 weeks after eggs are shed at temperatures above 22°C hatching can occur after 10 days.[48] Above 28°C most eggs die, but they can survive several months at low temperatures. So while egg shedding is continuous, increasing temperatures during early summer can give rise to simultaneous release of larvae from eggs that were shed when conditions were cold and eggs shed later when temperatures were warmer. This gives a rapid build-up of infectivity timed to coincide with the availability of snail intermediate hosts.[48]

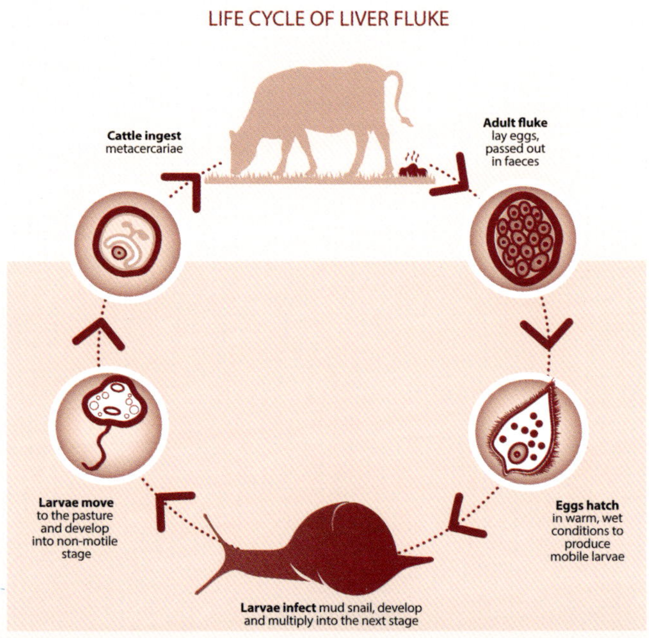

Fig. 37. Life cycle of liver fluke

The free-swimming larvae have 3 hours in which to find a suitable snail host in wet conditions. They penetrate the snail and multiply in its intestinal tract and liver. Several developmental stages occur inside the snail. The final stage larvae leave the snail and form infective cysts on appropriate plants. They are not released below 9°C or in the absence of free water but the onset of wet conditions, after a period of dry weather, can lead to the release of large numbers and rapid build up of infectivity for the final host.[48]

The encysted larvae are very hardy and can survive in dry and cold conditions but die if desiccated. The minimum period for completion of the lifecycle is 17 weeks, although in most cases takes 5–6 months.[48]

Pattern of disease

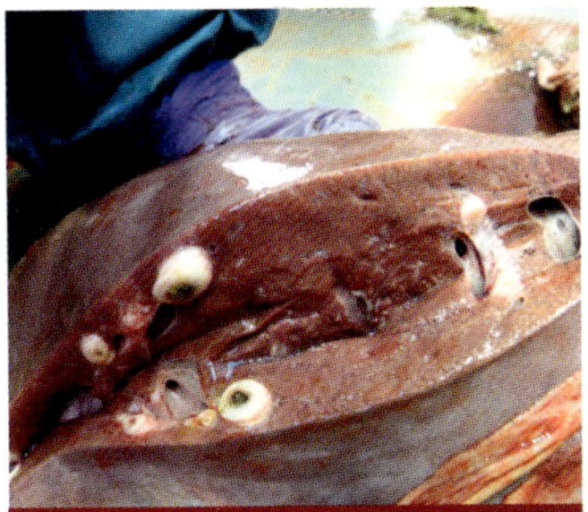

Fig. 38: "Pipe-stem liver" with calcified bile ducts. (Photo courtesy of M. Millar, School of Veterinary Science, University of Bristol)

Historically, temperatures in fluke areas in the west of the UK favoured snail development between May and October, the season being longer in the south and shorter in the north. Known as "Summer infection", snails were infected in mid-summer by larvae hatched from overwintered eggs, plus eggs shed in early spring and summer. The outcome of this infection being encysted larvae during the autumn.[48] Alternatively, encysted larvae can appear on pasture in May to June, derived from snails infected the previous autumn in which larval development had temporarily ceased during winter hibernation of the snail, and then recommenced in spring.[37] This is known as "Winter infection" of snails.

Incidence and geographic range of liver fluke has increased rapidly in the last 10 years. Milder, wetter winters, warmer summers and longer grazing seasons in all parts of the country, have meant enhanced overwinter survival of encysted larvae and snails on pasture, plus reduced time for fluke to reach egg-laying maturity. This has supported expansion of host snail and liver fluke populations.[47]

Additionally increases in animal movements, and environmental schemes aimed at preserving or retaining wetlands (favouring the mud snail) may have contributed to the spread of liver fluke. A reduction in farm labour may also have played a role with subsequent less attention paid to marginal areas (including drainage work and upkeep of fenced off boggy areas) commonly grazed by dairy dry cows and replacements.[50] There have also been increasing reports of flukicide resistance.

Several other important issues complicate the liver-fluke life cycle. It is possible to see immature fluke in the liver tissue and adult fluke in the bile ducts and gall bladder of the same animal at the same time, which suggests that cattle do not mount a strong protective immune response.[51] Liver fluke is not host-specific – any mammal that ingests plants with liver fluke cysts may become infected. This has important implications. First eradication is unlikely due to a significant wildlife reservoir; man can become infected (especially from the consumption of watercress from unfenced beds[37]) and cross-infection between cattle and sheep is possible, with obvious implications on treatment and management.[47]

Effects on the animal

Initially migration in liver tissue is associated with damage and haemorrhage. Then when adult fluke are in bile ducts, they feed on blood and damage the bile ducts themselves.[37]

Chronic liver-fluke infection in cattle causes anaemia and low blood protein levels. There is calcification of the bile ducts and enlargement of the gall bladder. The calcified bile ducts often protrude from the liver surface, giving rise to the term "pipe-stem liver"[37] (Fig. 38). Liver damage may also result in functional abnormalities. Aberrant migration of flukes does occur, with parasites seen in the lungs, and migration to the foetus has also been recorded, resulting in prenatal infection.[37]

Clinical signs

Acute and subacute disease, as seen in sheep, can occur in calves suffering heavy burdens of migrating immature fluke. Clinical signs are generally more obvious in sheep than in cattle, partly due to the size of cattle livers relative to the tissue damage caused, and partly due to the physical and physiological nature of cattle livers creating a less favourable environment for the migrating immature flukes.[48] Chronic disease is therefore far more significant in cattle.

Acute disease occurs 2–6 weeks after ingestion of large numbers of encysted larvae, and is due to the severe haemorrhage that results when the young flukes, migrating in liver tissue, rupture blood vessels. Damage to liver tissue is also severe.[37] Outbreaks of acute fluke are generally presented as sudden deaths during autumn and early winter. Examination of the rest of the herd may find some heifers that are weak, with pale membranes, have difficulty breathing and sometimes palpably enlarged livers with abdominal pain and fluid in the abdomen.[37]

Subacute disease occurs 6–10 weeks after ingestion of lower numbers of encysted larvae, and also appears in late autumn and winter. There is anaemia and lowered blood protein, and if untreated – high death rates. However, onset is not as rapid as acute disease and clinical signs are present for 1–2 weeks prior to death; these include rapid loss of condition, very pale membranes and an enlarged palpable liver. Swelling under the jaw or around the face and fluid in the abdomen may also be present.[37]

Chronic liver fluke disease is usually seen February to March onwards. Prolonged intake of low to moderate numbers of cysts from herbage results in progressive loss of body condition associated with the accumulation of adult flukes in bile ducts and gall bladder.[51] Severely affected heifers may become debilitated, with an increased incidence of metabolic and infectious diseases.[52] There is also the risk of prolonged calving to conception intervals and increased culling.

Diagnosis

Faecal egg counts (FEC) are the most frequently used method to diagnose liver fluke infection. A minimum of 6 animals should be sampled. An indication of herd level, but not an accurate indication of the number of fluke is possible, because adult fluke reside in the bile ducts, so eggs can become trapped there or in the gall bladder and only appear sporadically in the faeces.[51] It is also possible to have a positive FEC up to a few weeks after an animal has been successfully treated, or even have no eggs present when there is still an active fluke infection. This test also only detects the presence of egg-laying adults not juvenile fluke migrating through the liver.

Blood-enzyme profiles may be coupled with FEC. Alternatively, tests that detect anti-fluke antibodies in blood will indicate whether an animal has been exposed to infection. They are useful for surveillance, but not necessarily an indication of active infection as antibodies can persist many months after treatment.

Post-mortem examination of livers on-farm, the abattoir or Veterinary Investigation Centre is usually used to confirm disease. Chronic liver changes, the pipe-stem appearance of protruding bile ducts, plus the presence of adult fluke in the bile ducts and gall bladder are all used. Minimally invasive laparoscopic surgery has also been used to detect chronic liver disease in debilitated animals[53] (Fig. 39).

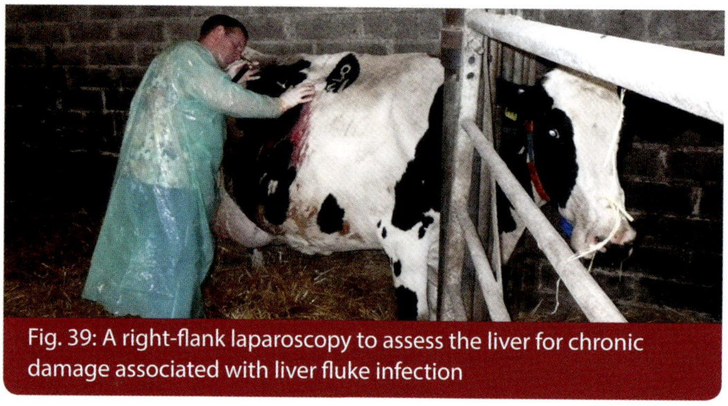

Fig. 39: A right-flank laparoscopy to assess the liver for chronic damage associated with liver fluke infection

Treatment

Only triclabendazole will remove early immature (around 2 weeks of age in cattle) stages of liver fluke.[37] Licensed treatments of chronic fluke infections are closantel, oxyclozanide, nitroxynil, albendazole and clorsulon.

A typical dosing schedule in an average rainfall season for heifers would be:[51]

- *Dose in spring (late April/May) to remove adult fluke burdens and reduce deposition of fluke eggs on pasture*
- *Dose in autumn (October) and again in winter (January) with a flukicide that is effective against immature fluke*
- *More regular treatments of stock may be necessary in high risk fluke areas or if the season has been very wet*
- *Stock bought in from areas with a high fluke incidence should be treated before joining the herd with a flukicide effective against immature stages.*

All heifers in the group should be dosed according to the heaviest individual. If there is wide variation in body weight then the group may be split and cattle dosed according to the heaviest in each sub-group.[54] Combination fluke and worm products should only be used when both groups of parasites are present as their use could potentially lead to off-target selection for resistance.[34]

None of the current flukicides has residual effects and animals are vulnerable to re-infection immediately after treatment.[47] Treated cattle should therefore be moved to clean pastures wherever possible. Improved nutrition of affected cattle is essential to restore body condition and production.[52]

Many of the available flukicides cannot be used in animals producing milk for human consumption, or have long milk withdrawal periods. This must always be borne in mind when treating pregnant heifers prior to their entry to the dairy herd.

Control

Control programmes must take into account farm history, topography, geographical location and prevailing weather. Most programmes rely heavily on flukicidal treatments. Choice of product and frequency of use depend on level of fluke challenge, time of year, and management and husbandry systems on the farm.[34] Herds should be monitored before a flukicide is used unless there is a history of fluke infection on the farm. Continued monitoring will determine the need for repeated reatments.[34]

The principal objective of a control programme is to minimise encysted larval challenge to susceptible final hosts. The complex life style of flukes affords different opportunities for such control:

- Use of flukicides to suppress fluke egg output and pasture larval challenge
- Drainage to reduce rate of egg hatching and survival of free-living stages of liver fluke and prevent establishment of adult snail populations
- Grazing management strategies to avoid contact between larvae and susceptible final hosts.

If sheep are present on a farm then fluke-control programmes in cattle must include sheep. Any sheep brought onto the farm should be treated following SCOPS guidelines as they may be a source of resistant fluke to all grazing animals.[34] Successful programmes depend on strategic application of all these methods.[55]

Presence of the snail intermediate host is integral to the life-cycle of liver fluke. As a result, methods aimed at reducing snail numbers or snail habitat, e.g. through effective drainage and/or reducing access of cattle to potential snail habitats during high risk periods by fencing off "fluky" pasture (Fig. 40) has been advocated as part of integrated fluke control.[47] Good drainage and maintenance of ditches can help reduce snail numbers, but cost may exceed the cost of fluke. Furthermore, when conditions are optimal, small numbers of snails yield as many larvae as large numbers in a crowded environment. Therefore, unless all snails and snail eggs on pastures grazed by cattle can be killed, snail control may be ineffective. Similar considerations applied to the use of molluscicides, which are now environmentally unacceptable.[55]

Fig. 40: Fencing off ditches would reduce cattle exposure to snail habitats

External parasites

Flies

Flies can be an intense source of irritation to cattle through their feeding activity, disease transmission and/or allergic reactions to their bites.

In the UK there are at least 20 common species of fly that feed on cattle.[56] There are blood-feeding flies, and those that feed on host secretions. Both sexes of blood-feeding flies such as stable flies, horseflies, blackflies and biting midges, take blood as a source of protein; in females, this is needed for ovarian development.[57]

Secretion-feeding flies feed on secretions from the host eyes, nose, body (sweat), udder, vagina or anus. Feeding activity of some secretion-feeding flies such as the face fly (*Musca autumnalis*) and head fly (*Hydrotea irritans*) can damage the eyes of cattle, allowing entry of pathogens, e.g. *Moraxella bovis* ("New Forest disease" or "pinkeye" (Fig. 41). *Hydrotea irritans* often clusters on the lower abdomen and udder, and the bacteria involved in summer mastitis have been isolated from these flies, indicating they may transmit the disease.[56]

Flies visiting cattle can also transmit disease, e.g. Bluetongue and BVD virus. Larger flies (e.g. horseflies), require a blood meal of about 0.5 ml. Interrupted feeding of these flies can therefore increase the risk of disease transmission as they move between several cattle, as well as increasing levels of irritation.[57] Fly activity may directly cause dramatic escape behaviour, in which self-injury can occur, or more commonly movement into shade, restlessness, skin rippling or simply stamping and tail switching. All these changes in behaviour result in reduced time spent grazing, resting and decreased performance.[56] Each horn fly takes 20 to 30 blood meals per day and more than 200 biting flies per heifer has been shown to have a significant economic impact on performance.[56] Infestation by fly larvae known as myiasis or "fly-strike" in wounds or skin lesions is a further risk.

Flies can be divided into three groups:

1. *Midges and mosquitoes*
2. *Horseflies*
3. *Blowflies and stable flies*

Some species produce only one generation per year, others produce several – the situation becomes intolerable with up to three or four generations per year, depending on the weather.

Control

Use insecticide impregnated ear tags (Fig. 42) or tail bands containing pyrethroids, together with either pyrethroid pour-on, spot-on or spray preparations for maximum effect against flies. Use of screens and electrocution traps for buildings also helps reduce fly nuisance. Manure heaps are a primary breeding site for many species of fly so farm hygiene is essential. Aerosol space sprays, residual insecticides applied to walls and ceilings and insecticide impregnated cards and strips may reduce fly numbers indoors. Insecticides may also be incorporated in solid or liquid fly baits, using attractants such as various sugary syrups or hydrolysed yeast and animal proteins.

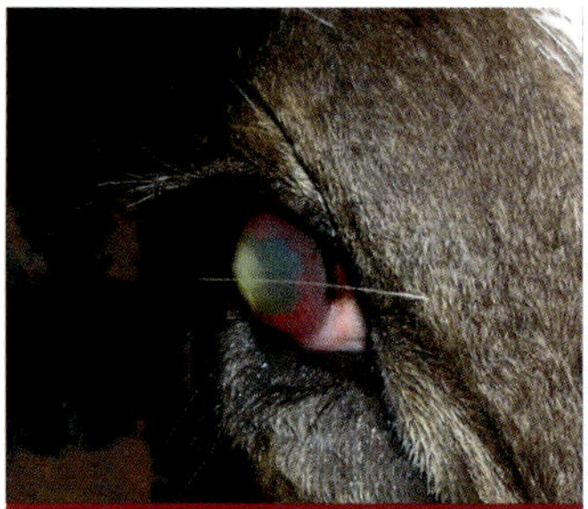

Fig. 41: New Forest disease or Pinkeye

(Photo courtesy of C. Penny)

Fig. 42: Pyrethroid impregnated eartag

Fly parasites can be purchased to tackle the fly before it becomes airborne. Fly parasites are parasitic wasps that burrow into pupae of un-hatched flies and eat them before they hatch, lowering fly numbers. The fly parasite larvae live for the whole fly season and can also overwinter. They should be sprinkled around the farm every fortnight – focusing on dirty areas – muck heaps, under water troughs, loose housing and any areas of standing water. Avoid areas where cows, tractors or people can crush them. Use fly parasites before any flies have had a chance to hatch out and breed.

Each bag of fly parasites contains around 50,000 pupae and one bag should be used for every 100 cows or heifers. Use the fly parasites along with spot-on or pour-on products for maximum reduction in fly numbers.

Lice

Lice are common parasites of UK cattle. All life-cycle stages are found simultaneously on the host. A nymph, who closely resembles the adult, hatches from an egg then moults several times until the adult stage is reached. Lice complete their entire life cycle on the animal, which takes about 3 weeks.

There are chewing lice and sucking lice. Chewing lice feed on skin and hair; sucking lice feed on blood using piercing mouthparts. The two are easily distinguished by the shape of their heads, and differentiation is important when selecting products for control.[56] Blood sucking lice have been implicated in the spread of tick-borne fever and ringworm, however, lice are predominantly of importance because of the direct damage they cause, which is usually a function of their density.[37]

Low burdens of lice are very common and often not a problem; lice being almost normal inhabitants of the skin and coat of cattle, especially in winter. However, louse populations can increase very rapidly. Moderate infestations are associated only with a mild chronic dermatitis and are usually well tolerated. In heavier infestations there is intense itching, with rubbing and licking. If sucking lice are present in large numbers there may be anaemia and weakness.[37] Heavy louse infestation is often a sign of an underlying condition, such as malnutrition or chronic disease, because debilitated animals may not groom themselves.

Transfer of lice between animals or herds is usually by direct physical contact. As lice do not survive for long off their host, (3–5 days depending on weather) the potential for animals to pick up infestations from dirty housing is limited.[56] Occasionally, lice may also be transferred between animals by attachment to flies.[37]

Lice and eggs can be found by parting the hair, especially along the midline. The lice are present next to the skin with the eggs scattered like coarse powder throughout the hair. Removal and examination under a light microscope will allow species identification.[37] In the UK, heaviest infestations are seen in late winter and early spring, when the coat is at its thickest, giving a sheltered, bulky and humid habitat for optimal multiplication. A rapid increase in louse populations is seen when cattle are housed for winter, and numbers build up quickly. In late spring, most of the parasites and eggs are shed with the winter coat, so numbers fall. They generally stay low throughout the summer, as the thinness of the coat provides a restricted habitat, and high skin surface temperatures plus direct sunlight limit multiplication and may even be lethal.[56]

Control

Identify and tackle any underlying disease issues as well as treating for lice. Pour-on or spot-on synthetic pyrethroids, such as deltamethrin, alpha-cypermethrin or permethrin are available for louse control, as are pour-on and injectable macrocyclic lactones (e.g. ivermectin, eprinomectin, moxidectin and doramectin)[58]. Injectables may have only limited activity against chewing lice and are more effective against sucking lice.

Most insecticides registered for use on cattle are not active against louse eggs. After treatment, therefore, eggs can still hatch and the newly hatched nymphs must be killed by the residual effects of the treatment. If, however, the residual efficacy of the product applied is short (less than 2 weeks) the newly hatched nymphs can continue the infestation.[58] In these cases, use a second treatment.

Timing and frequency of treatments will depend on individual circumstances. In many cases treatment in late autumn or early winter will give adequate control of cattle lice. All animals in the group must be treated. Undertake louse control when cattle are housed for the winter alongside treatment for other parasites. Treat all stock on farm and quarantine and treat all newly introduced animals to ensure effective louse control.[58]

Mites

Mites feed on blood, lymph, skin debris and secretions, which they ingest by puncturing the skin, scavenging from the skin surface or drinking from any lesions. Infestation by mites can result in severe dermatitis known as mange, which may cause significant welfare problems and economic losses.[37]

Eggs hatch into a 6 legged larva, which then moult through nymph and ultimately 8 legged adult stages. This may be completed in only 14 days. All life-cycle stages are found simultaneously on the host and they spend their whole lives there. Transmission from host to host is primarily by physical contact but also through contact with a contaminated environment (bedding, housing, trailers, etc).[56]

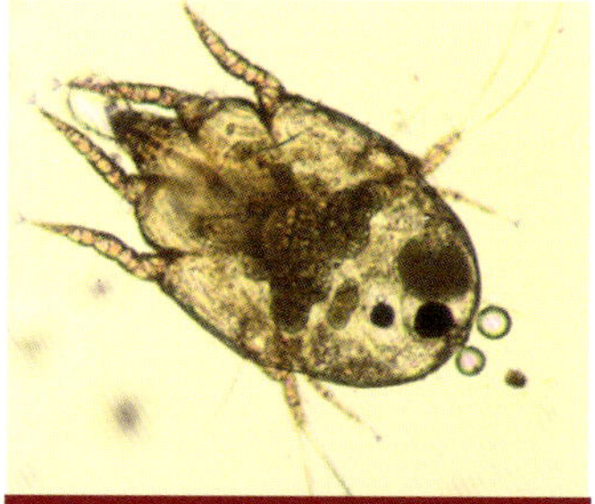

Fig. 43: *Chorioptes bovis* mite
(Photo courtesy of M. Millar, School of Veterinary Science, University of Bristol)

The commonest mite affecting UK cattle is Chorioptes bovis (Fig. 43). It occurs most frequently at housing and populations are highest in the winter. Chorioptic mange is most commonly seen on the feet, legs and base of the tail and udder (Fig. 44). It tends to only cause mild problems – lesions remain localised, with slow spread.[56]

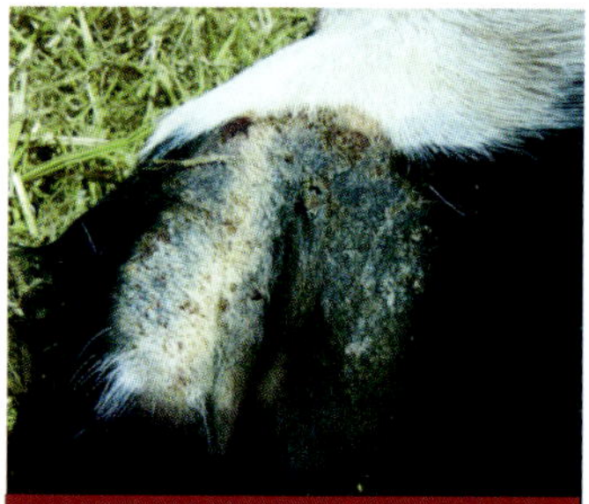

Fig. 44: *Chorioptes* mange on the tail head of a heifer
(Photo courtesy of M. Millar, School of Veterinary Science, University of Bristol)

Hosts can have no symptoms with low densities of mites, and act as carriers to transfer the mite to other animals. However, if mite numbers reach high densities clinical signs may be seen. Clinically affected animals may have pustular, crusted, scaly and thickened patches of skin with hair loss. This is usually confined to the tail head, legs and lower body but in some cases this may spread to other areas and cause disease.

The itching caused by the mites results in rubbing and scratching, with damage to the hide.[56]

Much less common than *Chorioptes bovis* are *Psoroptes bovis* and *Sarcoptes scabei*. Both these mites have potential to cause more severe signs than Chorioptic mange, and veterinary advice must be sought if they are suspected.

Control

Permethrin is the only pyrethroid with a claim in the UK against chorioptic and sarcoptic mange mites in cattle. Pour-on ivermectin, doramectin, eprinomectin and moxidectin are also effective.[56]

Treat all animals in a herd and any in-contact animals to eradicate the mites. Chorioptes mites are likely to survive for at least 3 weeks off the host, depending on temperature and humidity. So immediately remove treated animals to an area that has been free of potentially infested stock (particularly important when using products with low levels of residual activity).

Ticks

Ticks are blood-feeding parasites that are only present on the host during a short (several days) blood-feeding period. In the UK, intensities of infection in cattle are usually low, averaging between one and three ticks per infested animal.[56] Tick bites may be directly damaging to cattle, causing irritation, inflammation, hypersensitivity, susceptibility to secondary infections, and when present in large numbers, anaemia and production losses.

Fig. 45: The tick *Ixodes ricinus*

More importantly, when they attach and feed, ticks can transmit a number of infections, which in turn can lead to severe, sometimes fatal, disease.[59] The most important pathogens transmitted to cattle by ticks in the UK are *Babesia divergens* and *Anaplasma phagocytophilum*.

Redwater fever is caused by *Babesia divergens*; a protozoan parasite usually transmitted by the tick *Ixodes ricinus* (Fig. 45). The *Babesia* multiplies in red blood cells and ruptures them; this causes a fever, anaemia and the accumulation of haemoglobin (from the red blood cells) in the urine, hence "redwater". Pregnant animals can abort, and in high enough numbers *Babesia* causes death.

Calves up to 1 year old, although fully susceptible to infection, are relatively resistant to disease. Hence, in areas of high tick infection pressure, most animals become infected when young and acquire immunity without showing clinical signs. In older cattle, immunity is reinforced by repeated tick challenge. Thus, in areas where babesiosis is widespread, clinical cases are rare, although the parasite is detectable in most animals.[56]

Clinical babesiosis is most commonly seen when this stability breaks down, e.g. naive cattle introduced into an area of widespread babesiosis with high tick infestation pressure. As a result, babesiosis is often seen where one to two-year-old animals are introduced to marginal, tick-infested grazing.[56] Treatment involves a babesicide called imidocarb, supportive therapy such as blood transfusion and fluid therapy, and tick management.

Tick-borne fever is caused by *Anaplasma phagocytophilum* – a bacterium transmitted largely by *Ixodes ricinus*. The Anaplasma enters white blood cells and destroys them causing immunosuppression. There is a fever and increased risk of secondary infections such as pneumonia and Babesiosis. However, in the UK, infection with Anaplasma in cattle is usually mild, the main losses being abortion in pregnant animals.[56] Treatment is long-acting tetracycline antibiotics and tick management.

Control

Ticks spend most of their life cycle away from the host, sheltered at the base of thick damp vegetation, which makes control difficult. Ticks become active and start to feed in early spring, but start and duration of the tick season is difficult to predict because it is weather-dependent. Tick populations can be reduced by pasture improvement, drainage and scrub clearance, but this is a long-term exercise requiring sustained effort, and alternative hosts such as deer add further challenges![56]

If cattle are about to be moved to a known tick area, use prophylactic protection – pour on pyrethroids or macrocyclic lactones (neither has a UK label claim for cattle against ticks at present, so must be used under the cascade system). Reapply at regular intervals during the tick season to achieve sustained protection.[56]

References

1. Lorenz, I., Fagan, J. and More, S. J. (2011) Calf health from birth to weaning. II. Management of diarrhoea in pre-weaned calves. *Irish Veterinary Journal* 64
2. Aldridge, B. and Potter, T. (2011) A systematic approach to calf gastroenteric disease. *UK Vet* 16.
3. Stott, A. W. and Gunn, C. J. (1995) *Scottish Agriculture Economics Review* 8.
4. Andrews, A. H. (2004) Calf enteritis – diarrhoea in the pre-weaned calf – strategic investigation of outbreaks. *Cattle Practice* 12.
5. APHA (2013) Veterinary Investigation Diagnosis Analysis (VIDA) report, 2013. https://www.gov.uk/government/uploads/system/uploads/attachment_data/file/356653/vida-cattle-06-13.pdf. Accessed April 2016.
6. Mason, C., Caldow, G. and Katzer, F. (2012) Calf diarrhoea. *The Moredun Foundation News Sheet* Vol. 5 No. 14.
7. Head, M. F. (2008) The changing face of bovine Cryptosporidiosis. *UK Vet* 13 No. 8.
8. Muirhead, S. (1989) Coccidiosis infections often go undetected in beef and dairy cattle. *Feedstuffs* 15: 87.
9. Grove-White, D. (2010) Prevention and treatment of neonatal calf diarrhoea – a personal overview. *Cattle Practice* 18.
10. Lorenz, I. and Klee, W. (2007) Neonatal calf diarrhoea – something old, something new. *Cattle Practice* 15.
11. Lorenz, I. (2013) An update on research into factors affecting calf health. *Cattle Practice* 21.
12. Nappert, G. (2008) Review of current thinking on calf oral rehydration. *Cattle Practice* 16.
13. Smith, G. W. (2013) Antibiotic decision making – calf scours. *Proceedings of the 46th Annual Conference of the American Association of Bovine Practitioners*.
14. NAHMS Dairy 2007 http://www.aphis.usda.gov/animal_health/nahms/dairy/downloads/dairy07/Dairy07_ir_CalfHealth.pdf. Accessed April 2016
15. Andrews, A. H. (2000) Calf pneumonia costs! *Cattle Practice* 8.
16. Van Der Fels-Klerx, H. J., Saatkamp, H. W., Verhoeff, J. and Dijkhuizen, A. A. (2002) Effects of bovine respiratory disease on the productivity of dairy heifers quantified by experts. *Livestock Production Science* 75.
17. Esslemont, R.J., Kossaibati, M. A. and Reeve-Johnson, L. (1998) The costs of respiratory diseases in dairy heifers calves. *XX World Buiatrics Congress, Sydney*. (Proceedings, Volume 2, p685-691). C28. (75)
18. AHDB Dairy Website (2016) http://www.dairyco.org.uk/market-information/#.VXLJZNJViko. Accessed April 2016.
19. Waltner-Toews, D., Martin, S. W. and Meek, A. H. (1986) The effect of early calfhood health status on survivorship and age at first calving. *Canadian Journal of Veterinary Research* 50.
20. Warnick, L. D., Erb, H. N. and White, M. E. (1994) The association of calfhood morbidity with first-lactation calving age and dystocia in New York Holstein herds. *The Kenyan Veterinarian* 18.
21. Robertson, J. F. (2000) Building designs to optimise health. *Cattle Practice* 8.
22. Mason, C. (2011) Making the best use of diagnostics for cattle pneumonia. *Cattle Practice* 19.
23. Potter, T. (2015) Bovine respiratory disease. *Livestock* 20.
24. Gibbs, A. (2001) Practical approach to the control of pneumonia in housed calves. In Practice 23.
25. Lorenz, I., Earley, B., Gilmore, J., Hogan, I., Kennedy, E. and More, S. J. (2011) Calf health from birth to weaning. III. Housing and management of calf pneumonia. *Irish Veterinary Journal* 64:14.
26. Bach, A. (2011) Associations between several aspects of heifer development and dairy cow survivability to second lactation. *Journal of Dairy Science* 94.
27. Potter, T. (2010) Systematic approach to calf pneumonia. *UK Vet* 15.
28. Virtala, A. M., Mechor, G. D., Grohn Y. T. and Erb, H. N. (1996) The effect of calfhood diseases on growth of female dairy calves during the first 3 months of life in New York State. *Journal of Dairy Science* 79.

29. Blowey, R. W. (1999) *A Veterinary Book for Dairy Farmers* (3rd Edition). Farming Press.

30. Albright, J. L. (1987) Dairy animal welfare: current and needed research. *Journal of Dairy Science* 70.

31. Thomsett, L. R. (2004) Skin Conditions. In: Andrews, A. H. editor. *Bovine Medicine* (2nd Edition). Blackwell Publishing.

32. NOAH Compendium (2016) http://www.noahcompendium.co.uk/Elanco_Animal_Health/Imaverol_100_mg_ml_Concentrate_for_Cutaneous_Emulsion/-57356.html. Accessed May 2016.

33. NOAH Compendium (2016) http://www.noahcompendium.co.uk/MSD_Animal_Health/Bovilis_Ringvac/-28664.html. Accessed May 2016.

34. Taylor, M. A. (2010) COWS, Control of worms sustainably. *AHDB Technical Manual.*

35. Cattle Parasites (2014) http://www.cattleparasites.org.uk/guidance/manual/COWS%20Controlling%20parasitic%20gastroenteritis%20in%20cattle.pdf. Accessed May 2016.

36. Moredun Foundation (2016) http://www.moredun.org.uk/research/research-@-moredun/parasitic-worms/parasitic-control-cattle. Accessed May 2016.

37. Taylor, M. A., Coop, R. L. and Wall, R. L. (2016) Parasites of cattle. In: *Veterinary Parasitology* (4th Edition). Wiley, Blackwell Publishing.

38. AHDB Beef and Lamb (2016) The BRP Cattle and Sheep Parasite Control Product Guide. Better Returns Programme.

39. Cattle Parasites (2014) http://www.cattleparasites.org.uk/guidance/manual/COWS%20Integrated%20parasite%20control%20on%20cattle%20farms.pdf. Accessed May 2016.

40. Cattle Parasites (2014) http://www.cattleparasites.org.uk/guidance/manual/COWS%20Controlling%20lungworm%20in%20cattle.pdf. Accessed June 2016.

41. David, G. P. (1999) Strategies for the control of parasitic bronchitis in cattle. In Practice 21.

42. Matthews, J. (2008) Clinical forum: bovine lungworm. *UK Vet* 13.

43. Intervet UK Limited. (2001) Strategies for the effective control of lungworm. *Bovilis Huskvac Technical Manual.*

44. David, G. P. (1997) An epidemiological study of husk in adult cows in 32 UK herds (preliminary findings). *Cattle Practice* 5.

45. Matthews, J. (2004) Lungworm Disease in Cattle. *The Moredun Foundation News Sheet* Vol. 4.

46. Schweizer, G., Braun, U., Deplazes, P. and Torgerson, P. R. (2005) Estimating the financial losses due to bovine fasciolosis in Switzerland. *Veterinary Record* 157.

47. Skuce, P. J. and Zadoks, R. N. (2013) Liver fluke – a growing threat to UK livestock production. *Cattle Practice* 21.

48. Sargison, N. (2008) Fluke diseases of UK ruminant livestock Part 1: Life cycles, economic consequences and diagnosis. *UK Vet* 13 No. 5.

49. Forbes, A. (2013) Liver fluke control in cattle: why, when and how? *Cattle Practice* 21.

50. Sargison, N. (2010) Clinical Forum: Fasciolosis in sheep and cattle. *UK Vet* 15.

51. Skuce, P. J. and Zadoks, R. N. (2013) Fluke control in sheep and cattle. *The Moredun Foundation News Sheet* Vol. 5.

52. Scott, P. (2009) Disease facts – chronic fasciolosis (liver fluke). *UK Vet* 14.

53. Tilling, O. (2014) Liver fluke in cattle in the United Kingdom. *Livestock* 19.

54. Dobbs, M. (2014) Managing liver fluke in cattle and sheep, whatever the weather. *Veterinary Times* 44.

55. Sargison, N. (2008) Fluke diseases of UK ruminant livestock Part 2: Treatment and control. *UK Vet* 13.

56. Cattle Parasites (2014) http://www.cattleparasites.org.uk/guidance/manual/COWS%20Controlling%20ectoparasites%20and%20insect%20pests%20of%20cattle.pdf. Accessed June 2016

57. Rankin, M. (1998) An Overview of Nuisance Flies on Cattle. *Cattle Practice* 6.

58. Foster, A., Mitchell, S. and Wall, R. (2015) Cattle ectoparasites in Great Britain. *Cattle Practice* 23.

59. Taylor, M. A., Coop, R. L. and Wall, R. L. (2016) Veterinary Entomology. *Veterinary Parasitology* (4th Edition). Wiley, Blackwell Publishing.

60. www.vetmed.wisc.edu/dms/fapm/fapmtools/8calf/calf_health_scoring_chart.pdf

Index

Veterinary Guide To Rearing Dairy Heifers

Index